SpringerBriefs in Food, Health and Nutrition Series

Springer Briefs in Food, Health, and Nutrition present concise summaries of cutting edge research and practical applications across a wide range of topics related to the field of food science.

For further volumes:
http://www.springer.com/series/10203

SpringerBriefs in Food, Health and Nutrition Series

[illegible]

Editor-in-Chief

[illegible]

Associate Editors

[illegible]

Fidel Toldrá • Milagro Reig

Analytical Tools for Assessing the Chemical Safety of Meat and Poultry

Fidel Toldrá
Instituto de Agroquímica y Tecnología
de Alimentos (CSIC)
Avenue Agustín Escardino 7
Paterna (Valencia), Spain

Milagro Reig
Institute of Food Engineering
for Development
Universidad Politécnica de Valencia
Camino de Vera s/n
Valencia, Spain

ISBN 978-1-4614-4276-9 ISBN 978-1-4614-4277-6 (eBook)
DOI 10.1007/978-1-4614-4277-6
Springer New York Heidelberg Dordrecht London

Library of Congress Control Number: 2012941363

Printed on acid-free paper

Springer is part of Springer Science+Business Media (www.springer.com)

Acknowledgements

Grants Prometeo/2012/001 from Conselleria d'Educació, Formació I Ocupació, as well as A-05/08 and A01/09 from CSISP (Food Safety Research) of Conselleria de Sanitat, from Generalitat Valenciana (Spain), are fully acknowledged. Work was performed under the Associated Unit IAD (UPV)-IATA (CSIC).

Contents

Chapter 1
Analytical Tools for Assessing the Chemical Safety of Meat and Poultry

1.1 Introduction

Meat and poultry are foods that contain important nutrients like high-biological-value proteins, group B vitamins, minerals and trace elements, and other bioactive compounds. Despite these benefits, the image of these meats for consumers is negative, especially red meats, because of the content of saturated fatty acids, cholesterol, and other substances that may contribute to a higher risk of contracting certain diseases. In fact, recent metastudies involving large numbers of volunteers suggest a relation between meat consumption or dietary heme and risk of colon cancer (Cross et al. 2010; Santarelli et al. 2010; Bastide et al. 2011) or even cardiovascular diseases and diabetes mellitus (Micha et al. 2010). Diets associated with cooked or cured meats have also shown an incidence of human cancers (Jaksyn et al. 2004). Consumer health and well-being are of outmost importance for international agencies and industry worldwide. This fact has driven relevant food research efforts toward strategies designed to improve the nutritional properties of meat and poultry by reducing the content of unhealthy substances and promoting the presence of other substances with healthy benefits (Toldrá and Reig 2011). In this way, the development of modern analytical technologies linked to epidemiologic studies and research conducted on the safety aspects of food have contributed to the detection of a large number of substances in food at very small amounts. These substances may be quite varied in nature, and their presence may be due to different reasons or causes; sometimes they are deliberately added for increased profitability, while in other cases they are accidentally generated in certain processing conditions. Some of these substances have shown relevant toxic consequences for consumers like carcinogenicity, genotoxicity, or other undesirable effects on human health, and thus they must be controlled to assure consumer safety.

This manuscript has been divided into two large groups of substances: (1) those substances like growth promoters, antibiotics, carcass disinfectants, and environmental contaminants that may be present, either incidentally or deliberately, in raw

F. Toldrá and M. Reig, *Analytical Tools for Assessing the Chemical Safety of Meat and Poultry*, SpringerBriefs in Food, Health, and Nutrition 9,
DOI 10.1007/978-1-4614-4277-6_1,

meat and poultry and (2) substances that may be generated during further processing of meat and poultry like N-nitrosamines (generated when using nitrite as a preservative under certain processing conditions), polycyclic aromatic hydrocarbons (PAHs) (generated in certain smoking processes), heterocyclic amines (generated when cooking at high temperatures), biogenic amines (generated by microbial decarboxylation of certain amino acids), certain oxides (generated from protein or lipid oxidation), and radiolytic products (generated when performing irradiation). Thus, the manuscript provides a review of how such groups of residues could be either present in meat or poultry or generated as a consequence of further processing. It also discusses their health-related effects for consumers and the available analytical tools for their detection and control.

1.2 Control Tools to Assure the Chemical Safety of Meat and Poultry and Derived Products

Safety is an important issue in global commercial food transactions; this is especially relevant for the meat-processing industries where globalization implies a large volume of raw-material and final-product exchanges among countries. Controls and effective corrective measures are basic to assuring consumer safety. Thus the meat and poultry industries must implement adequate control systems to guarantee the safety of their supplies and final products and to comply with legislative requirements (Toldrá 2004).

The safety of processed meat or poultry depends on many factors including the initial raw materials, ingredients and additives, processing conditions like fermentation, drying, cooking or ripening, the type of packaging, and, finally, the storage conditions within the same industry and during commercial distribution (Toldrá 2006a; Reig and Toldrá 2007). It is thus necessary to control the absence of harmful substances, through the use of analytical methodologies that will be described later on in this manuscript, at all stages: raw materials, processing, and final product (Toldrá 2006b). The groups of substances that may be present in either raw meat and poultry or their derived products are summarized in Table 1.1, and the types of controls at each stage are compiled in Table 1.2.

1.2.1 Control of Raw Meats and Poultry

The control of raw materials is essential so that any meat showing the presence of a given residue that may be harmful to humans may be discarded. At first, residues suspected of being present in lean meat or poultry would be growth promoters and antibiotics, substances that might have been used on farms during animal production. Another relevant group of substances that may be incidentally present

Table 1.1 Groups of substances that need to be controlled in either raw meat and poultry or in processed meats and poultry

Group of substances	Type of food
Growth promoters	Raw meats and poultry
Veterinary drugs	Raw meats and poultry
Environmental contaminants	Raw meats and poultry
Carcass disinfectants	Raw meats and poultry
Nitrosamines	Cured meats
Biogenic amines	Fermented sausages
Heterocyclic amines	Cooked meats at high temperature
Polycyclic aromatic hydrocarbons	Smoked meats and poultry
Lipid oxidation products	Processed meats and poultry
Protein oxidation products	Processed meats and poultry
Radiolytic products	Irradiated meat and poultry

Table 1.2 Safety controls to be considered at each processing stage

Stage	Location products	Controls
Raw materials	Reception	Hygiene
	Lean meat and poultry	Presence of growth promoters, veterinary drugs, environmental contaminants, or disinfectants
		Oxidation of proteins
	Reception	Hygiene
	Fat	Presence of environmental contaminants or radiolytic products
		Oxidation of lipids
Process: fermentation	Curing chamber	Microbial growth
	Fermented sausages	Generation of amines
Process: drying	Curing chamber	Microbial growth, presence of molds, microbial or mold metabolites
	Dry sausages, dry-cured ham	Generation of nitrosamines or amines
		Oxidation of proteins and lipids
Process: smoking	Smoking chamber	Generation of polycyclic aromatic hydrocarbons
	Smoked meats and poultry	
Process: cooking	Cooking/frying	Generation of heterocyclic amines
	Cooked meats and poultry	Oxidation of proteins and lipids

are environmental substances due to the use of contaminated ingredients in the feed used for animal production. The fat must also be analyzed for the detection of fat-soluble substances. In some countries, beef, pig, and poultry carcasses may be externally treated, through spray or immersion, with some food disinfectants, either prechill or postchill. Depending on the nature of the disinfectant, it may remain in either the lean tissue or the fat. In all such cases, analytical determinations must be performed with groups of growth promoters, antibiotics, environmental substances, and disinfectants to assure the absence of any of these chemicals in the meat (Table 1.2).

1.2.2 Controls During Processing

In addition to substances that may be present in raw meat or poultry, several groups of substances may be generated as a consequence of processing and thus must be controlled or prevented (Reig and Toldrá 2010). The stage, the location in the factory for sampling, and the controls to be performed are given in Table 1.2.

The generation of nitrosamines may be prevented through the use of legally permitted nitrites (i.e., 125 ppm in the USA and 150 ppm in the EU) and assuring that low amounts of residual nitrite are left, just to minimize the possibility of interaction with secondary amines (Pegg and Shahidi 2000). A valid alternative is the addition of ascorbic acid, which would ensure the rapid reduction of nitrite into nitric oxide, avoiding any residual nitrite (Cassens 1997). The generation of biogenic amines is due to the action of microbial decarboxylase (Toldrá 2004). The best way to prevent amines is to control microbial starters used in meat fermentation, verifying the absence of such decarboxylase activity (Toldrá 2006a, b). Heterocyclic amines are generated at high cooking temperatures so that their formation may be prevented or at least reduced by controlling the cooking conditions. Oxidation of proteins and lipids may be prevented through the use of adequate antioxidants (Estévez et al. 2009). In the case of smoke flavorings, preventive measures include the use of correctly produced primary products and the control of PAHs (Simko 2009a). In general, all these preventive measures are easy to implement in the industry and contribute to minimizing the problem in cooked, cured, and dry-cured meat products.

1.2.3 Controls in the Final Product

Once the products have already been produced and are ready for distribution and sale, several important controls must be performed to verify their final safety. The most important ones are given in Tables 1.1 and 1.2 and are briefly described in this manuscript.

1.3 Veterinary Drugs

Veterinary pharmaceutical drugs have long been used in animal production as therapeutic agents to control infectious diseases or as prophylactic agents to prevent outbreaks of diseases and control parasitic infections (Dixon 2001). Some of these drugs, like anabolic agents, may produce certain growth-promoting effects and improve the feed conversion efficiency, and they also increase the lean-to-fat ratio with a clear benefit to farmers. The weight increase is due partly to an inhibitory effect on muscle proteases (Fiems et al. 1990) and partly to increased fat utilization (Brockman and Laarveld 1986). The resulting meat is leaner (Lone 1997) but tougher because of the accumulation of connective tissue and collagen crosslinks (Miller et al. 1989, 1990). Meat products may also contain different types of residues having their origins in the meat used as raw material (Reig and Toldrá 2007). In addition, there are some potential adverse health effects (genotoxic, immunotoxic, carcinogenic, or endocrine) if animal tissues containing such residues are consumed. Other drugs such as antimicrobial agents have been used because they increase the availability of nutrients to the animal and improve the efficiency in the feed conversion rate. Fraudulent practices consist in using mixtures of several substances at very low amounts to obtain a synergistic effect for growth promotion (Monsón et al. 2007), making their detection by official control laboratories rather difficult (Reig and Toldrá 2007).

The differences in the national maximum residue limits (MRLs) are primarily attributable to differences in the level of risk that individual governments are prepared to accept, methodologies for establishing MRLs, and the conditions of use described in labeling of products (Reeves 2010). The existence of differing national standards adversely affects international trade in animal-derived food commodities by requiring exporters to comply with a diverse range of standards imposed unilaterally by importing countries.

1.3.1 *Causes of Concern for the Presence of Veterinary Drug Residues in Meat and Poultry*

Most veterinary drugs are orally active substances and can be administered either in feed or in drinking water. In some cases, such as active hormones, they are administered through implants in the subcutaneous tissue of the ears for slow release. The amount of residues in the injection sites is large, making withdrawal periods much longer (Reeves 2007). The main problem is that these substances or their metabolites may remain in meat and other foods of animal origin and may cause adverse effects on consumer health.

The European Food Safety Authority (EFSA) recently issued an opinion about the contribution of residues in meat and meat products of substances with hormonal activity, specifically testosterone, trenbolone acetate, zeranol, and melengestrol acetate (European Food Safety Authority 2007), but a quantitative estimation of risk to consumers could not be established.

Diethylestilbestrol is perhaps the most well-known substance since the connection between its genotoxic and mutagenic effects and cancer had been established in the 1940s (Lone 1997). Zeranol is a potent estrogen receptor agonist (Takemura et al. 2007), resembling estradiol in its action (Leffers et al. 2001). β-agonists may cause serious effects on consumers as observed in Italy following the consumption of clenbuterol in lamb and bovine meat. Effects include gross tremors of the extremities, tachycardia, nausea, headaches, and dizziness (Barbosa et al. 2005).

In the last decade, the abuse of antibiotics in farm animals has been the cause of great concern because of the development of increased bacterial resistance to certain antibiotics (Butaye et al. 2001), which were recently banned (Reig and Toldrá 2009a). Many antibiotics like chloramphenicol, nitrofurans, enrofloxacin, or sulphonamides are typically used for growth promotion practices that can create adverse effects on human health (Reig and Toldrá 2007). For instance, chloramphenicol may cause an irreversible type of bone marrow depression that could lead to aplastic anemia (Mottier et al. 2003), sulphonamides may be toxic to the thyroid gland (Pecorelli et al. 2004), and enrofloxacin may cause certain allergic reactions as well as the emergence of drug-resistant bacteria (Cinquina et al. 2003). Furazolidone, a metabolite of nitrofuran, has been reported as having mutagenic and carcinogenic properties (Guo et al. 2003), and sulfamethazine has been reported to contribute to tumor production. Coccidiostat residues may be present in poultry products treated with anticoccidials to prevent and control coccidiosis (Hagren et al. 2005), but they produce toxic effects on humans such as the dilatation of coronary arteries (Peippo et al. 2005).

Another relevant and disturbing negative effect is the potential development of resistant bacteria in the gastrointestinal tract (Butaye et al. 2001). The presence of antibiotics in meat may alter intestinal microflora (Chadwick et al. 1992; Vollard and Clasener 1994), which are subject to large variations in the proportion of major bacterial species (Moore and Moore 1995), or even disrupt the colonization barrier of the resident intestinal microflora (Cerniglia and Kotarski 2005), increasing their susceptibility to infection by pathogenic microorganisms (Cerniglia and Kotarski 1998). Furthermore, vancomycin-resistant enterococci, present as a consequence of the use of avoparcin, have been found in the commensal flora of farm animals, on meat from these animals, and in the commensal flora of healthy humans (van den Bogaard et al. 2000). Increased susceptibility to infection by pathogens like *Salmonella* spp. and *Escherichia coli* could be another indirect effect of this resistance (Cerniglia and Kotarski 1998).

1.3.2 Growth Promoters

Several groups of substances may be used to promote growth. The most common ones are briefly summarized below:

Steroid hormones and other substances having hormonal action. These substances exert estrogenic (except 17β-estradiol and ester-like derivatives), androgenic, or gestagenic action and may be used to promote growth (Table 1.3). Steroid hormones are essential for the normal development and physiological function of most tissues. Synthetic hormones may to bind to steroid receptors with equal or higher affinity than natural hormones (Wilson et al. 2002; Perry et al. 2005). Thus, trenbolone mainly binds to the androgen receptor and zeranol to the estrogen receptor, whereas

Table 1.3 Main properties of relevant androgens, estrogens, and gestagens (Reig and Toldrá 2009a)

Substance	IUPAC name	CAS number	Structure	Formula	Molecular mass (g/mol)	Melting point (°C)	Solubility in water (g/mL)
Androgens:							
17α-nortestosterone	17α-hydroxyestr-4-en-3-one		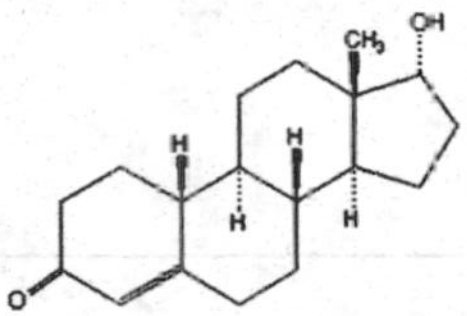	$C_{18}H_{26}O_2$	274.39	156	≈ Insoluble
17α-trenbolone	17α-hydroxyestra-4,9,11-trien-3-one	10161-33-8	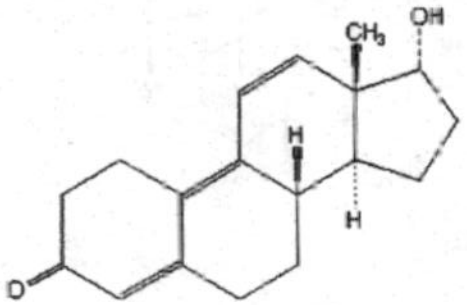	$C_{18}H_{22}O_2$	270.38	186	≈ Insoluble
17-methyltestosterone	(17β)-17-hydroxy-17-methylandrost-4-en-3-one	58-18-4	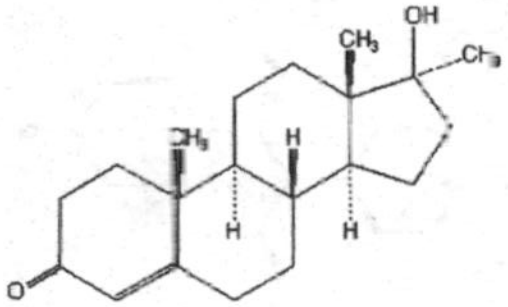	$C_{20}H_{30}O_2$	302.44	151	≈ Insoluble
Estrogens:							
Diethylstilbestrol	4,4′-(1,2-diethyl-1,2-ethene-diyl) bisphenol; α,α'-diethylstilbenediol	56-53-1	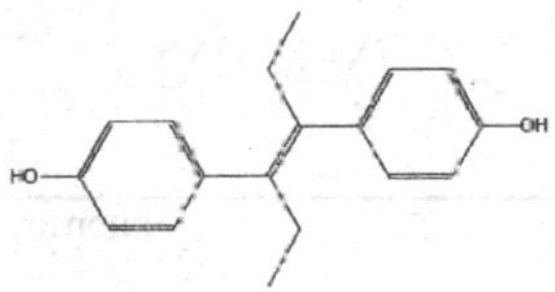	$C_{18}H_{20}O_2$	268.34	169	≈ Insoluble

(continued)

Table 1.3 (continued)

Substance	IUPAC name	CAS number	Structure	Formula	Molecular mass (g/mol)	Melting point (°C)	Solubility in water (g/mL)
Dimestrol	(E)-1,1′-(1,2-diethyl-1,2-ethenediyl)bis [4-methoxybenzene]; α,α'-diethyl-4,4′-dimethoxystilbene	130-79-0		$C_{20}H_{24}O_2$	296.39	124	≈ Insoluble
Dienestrol	4,4′-(1,2-diethylidene-1,2-ethanediyl) bisphenol; 4,4′-(diethylideneethylene) diphenol	84-17-3		$C_{18}H_{18}O_2$	266.32	227	≈ Insoluble
Gestagens:							
17α-hydroxy-progesterone	17-hydroxypregn-4-ene-3,20-dione	68-96-2		$C_{21}H_{30}O_3$	330.45	222	≈ Insoluble

melengestrol resembles natural progestins (EFSA 2007). MRLs have been established by national authorities and by the Codex Alimentarius. An important challenge when analyzing these residues in meat is the ability to discriminate between endogenous production and exogenous administration.

Stilbenes. These substances are synthetic nonsteroidal estrogens. They exert estrogenic activity (growth and development of female sexual organs) and produce an increase of somatotropin secretion. Diethylestilbestrol was related to cancer and is banned because it leads to several reactive metabolites after oxidation in the body (Lone 1997). Other stilbenes belonging to this group and its main properties are shown in Table 1.3.

Antithyroid agents. These agents are able to interfere directly or indirectly in the synthesis, release, or effect of thyroid hormones. These agents cause hypothyroidism by decreasing the basal metabolic rate, enlarging water retention, and thereby increasing the weight. Representative compounds and their main properties are shown in Table 1.4.

Glucocorticoids. Corticoids are hormones of the adrenal cortex that have physiological roles like the control of mineral and water balance. Glucocorticoids also have many important physiological functions like carbohydrate metabolism. They are used as anti-inflammatory agents for therapeutic purposes. Derivatives of prednisolone constitute the most important group of synthetic corticoids. Corticoids may exert some growth promotion when used in combination with other hormones or β-agonists. Corticoids used for such purposes include dexamethasone, betamethasone, flumethasone, cortisone, desoxymethasone, and hydrocortisone. Their main properties are given in Table 1.5.

β-agonists. β-adrenergic agonists are used as therapeutic agents for respiratory disorders by prescription of veterinary inspectors. However, they have been extensively used as growth promoters because they bind to the β receptors of various tissues and change the carcass composition. These substances reduce proteolysis and increase protein synthesis and lipolysis (Lone 1997). The group includes numerous substances such as, for example, clenbuterol, mabuterol, cimaterol, and salbutamol. Table 1.6 presents the group's main properties.

1.3.3 Antimicrobial and Antibiotic Drugs

Sulfonamides. This family of drugs is derived from sulfanilamide. Representative compounds are presented in Table 1.7. They are broad-spectrum antibiotics that are active against gram-positive and gram-negative bacteria, acting on specific targets in bacterial DNA synthesis (Croubels et al. 2004), and have been used in human medicine for the treatment of systemic bacterial diseases, although they have been replaced by modern antibiotics. Some of them, like sulfamethazine (also known as sulfamidicine), are still used in animals due to their low cost, easy administration, and high efficiency (Dixon 2001).

Table 1.4 Main properties of relevant antithyroideal agents (Reig and Toldrá 2009a)

Substance	IUPAC name	CAS number	Structure	Formula	Molecular mass (g/mol)	Melting point (ºC)	Solubility in water (g/mL)
Methylthiouracil	2,3-dihydro-6-methyl-2-thioxo-4(1 H)-pyrimidinone	56-04-2	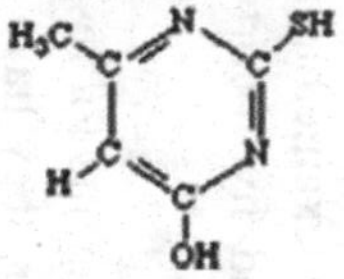	$C_5H_6N_2OS$	142.18	326	Slightly soluble (1:150 boiling water)
Propylthiouracil	2,3-dihydro-6-propyl-2-thioxo-4(1 H)-pyrimidinone	51-52-5	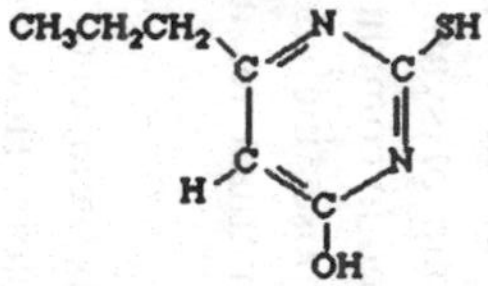	$C_7H_{10}N_2OS$	170.23	219	Slightly soluble (1:900)
Tapazole	1,3-dihydro-1-methyl-2 H-imidazole-2-thione	60-56-0	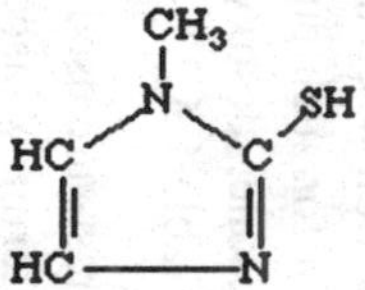	$C_4H_6N_2S$	114.17	146	Freely soluble
Thiouracil	2,3-dihydro-2-thioxo-4(1 H)-pyrimidinone	141-90-2	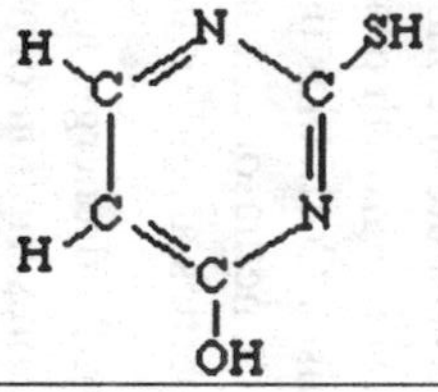	$C_4H_4N_2OS$	128.15	No definite	Very slightly soluble (1:2000)

Table 1.5 Main properties of relevant glucocorticoids (Reig and Toldrá 2009a)

Substance	IUPAC name	CAS number	Structure	Formula	Molecular mass (g/mol)	Melting point (°C)	Solubility in water (g/mL)
Betamethasone	9-fluoro-11,17,21-trihydroxy-16-methylpregna-1,4-diene-3,20-dione	378-44-9	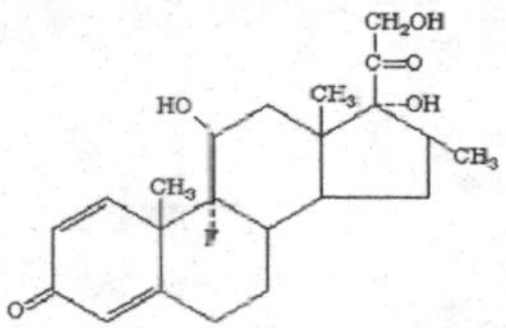	$C_{22}H_{29}FO_5$	392.45	231	–
Dexamethasone	(11β,16α)-9-fluoro-11,17-21-trihydroxy-16-methylpregna-1,4-diene-3,20-dione	50-02-2	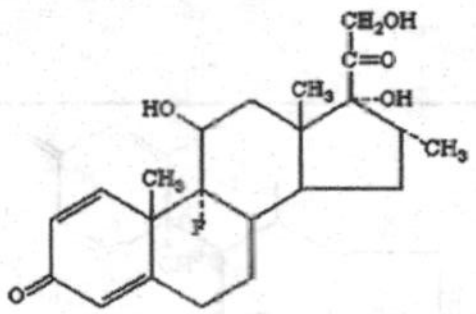	$C_{22}H_{29}FO_5$	392.45	268	Slightly soluble (0.01)
Flumethasone	6,9-difluoro-11,17,21-trihydroxy-16-methylpregna-1,4-diene-3,20-dione	2135-17-3	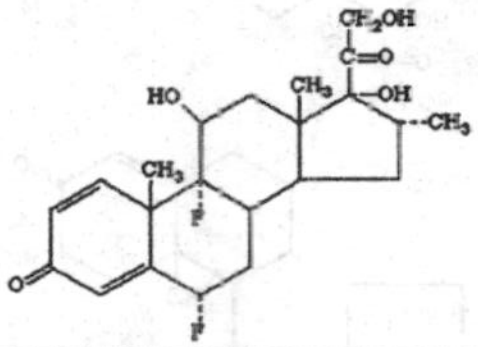	$C_{22}H_{23}F_2O_5$	410.46	260	Insoluble

(continued)

Table 1.5 (continued)

Substance	IUPAC name	CAS number	Structure	Formula	Molecular mass (g/mol)	Melting point (°C)	Solubility in water (g/mL)
Corticosterone	(11β)-11,21-dihydroxypregna-4-ene-3,20-dione	50-22-6	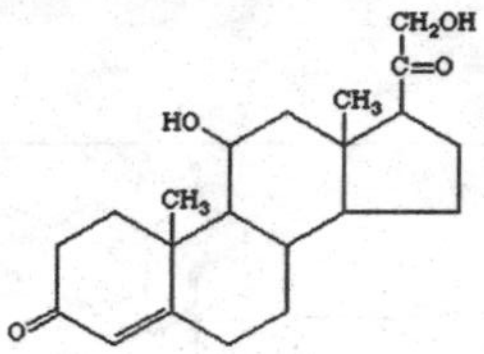	$C_{21}H_{30}O_4$	346.45	180	Insoluble
Cortisone	17α,21-dihydroxy-4-pregnene-3,11,20-trione	53-06-5	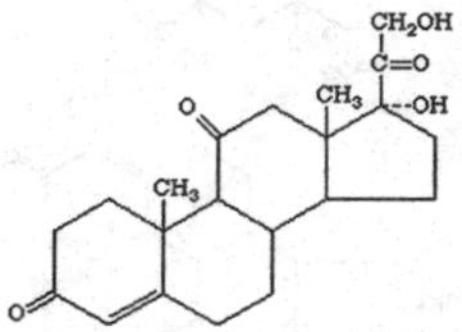	$C_{21}H_{28}O_5$	360.46	220	Slightly soluble (0.028)

Table 1.6 Names and main properties of representative agonists (Reig and Toldrá 2009a)

Substance	IUPAC name	CAS number	Structure	Formula	Molecular mass (g/mol)	Melting point (°C)
Clenbuterol	4-amino-α-[(tert-butilamino) methyl]-3,5-dichlorobenzyl alcohol	37148-27-9	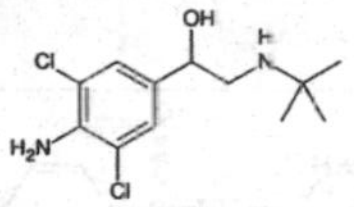	$C_{12}H_{18}N_2Cl_2O$	277.19	174
Mabuterol	4-amino-3-chloro-α-[(1,1-dimethyl-ethyl) amino]methyl]-(5-trifluoromethyl) benzenemethanol	56341-08-3	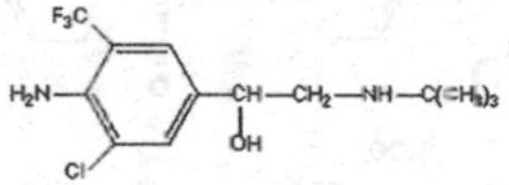	$C_{13}H_{18}N_2F3ClO$	310.75	205
Salbutamol	2-(hidroximetil)-4-[1-hidroxi-2-(tert-butilamino)etil]fenol	18559-94-9	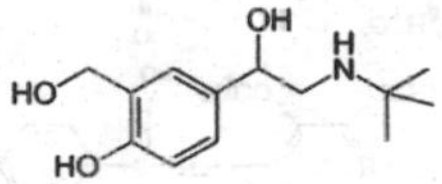	$C_{13}H_{21}NO_3$	239.31	157
Cimaterol	2-amino-5-[1-hydroxy-2-[(1-methyl-ethyl) amino]ethyl]benzonitrile	54239-37-1	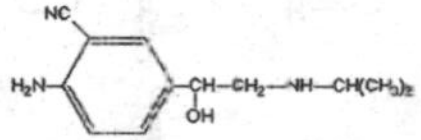	$C_{12}H_{17}N_3O$	219.29	159
Brombuterol	1-(4-Amino-3,5-dibromophenyl) -2-*tert*-butylaminoethanol	21912-49-2	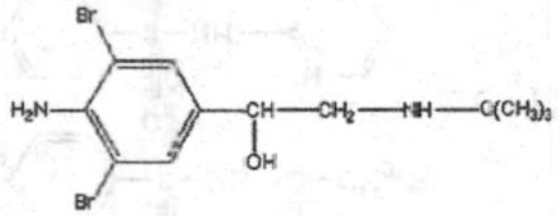	$C_{12}H_{18}Br_2N_2O$	366.08	—
Mapenterol	1-(4-Amino-3-chloro-5-trifluoromethylphenyl) -2-(1,1-dimethylpropylamino) ethanol	54238-51-6	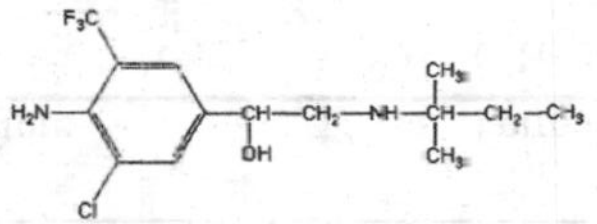	$C_{14}H_{20}ClF_3N_2O$	324.76	165
Ractopamine	4-hydroxy-alpha-[[[3-(4-hydroxyphenyl)-1-methylpropyl]amino]methyl] benzenemethanol	97825-25-7	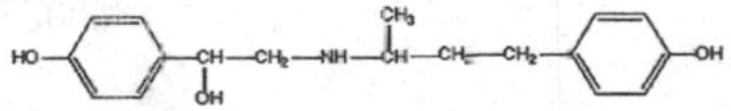	$C_{18}H_{23}NO_3$	301.39	124

Table 1.7 Main properties of relevant sulfonamides (Reig and Toldrá 2009a)

Substance	IUPAC name	CAS number	Structure	Formula	Molecular mass (g/mol)	Melting point (°C)	Solubility in water (g/mL)
Sulfacetamide	N-[(4-aminophenyl) sulfonyl]-acetamide	144-80-9	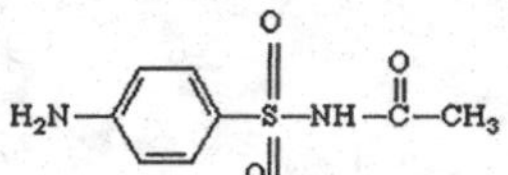	$C_8H_{10}N_2O_3S$	214.24	182	Slightly soluble (1:150)
Sulfadiazine	4-amino-N-2-pyrimidinylsul-fanilamide	68-35-9	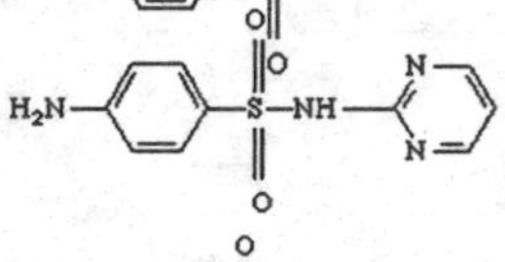	$C_{10}H_{10}N_4O_2S$	250.28	252	Slightly soluble in warm water
Sulfadoxine	4-amino-N-(5,6-dimethoxy-4-pyrimidinyl) benzenesulfonamide	2447-57-6	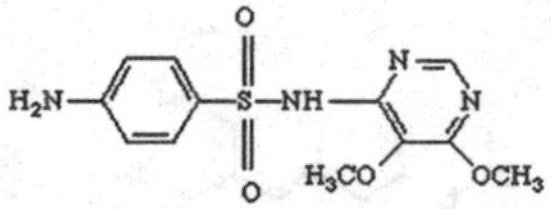	$C_{12}H_{14}N_4O_4S$	310.34	190	Slightly soluble
Sulfadimethoxine	4-amino-N-(2,6-dimethoxy-4-pyrimidinyl) benzenesulfonamide	122-11-2	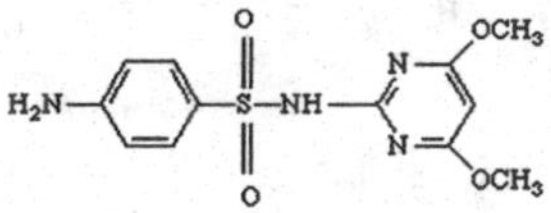	$C_{12}H_{14}N_4O_4S$	310.33	201	Soluble in slight acid solutions
Sulfachlorpyridazine	4-amino-N-(6-chloro-3-pyridazinyl) benzenesulfonamide	80-32-0	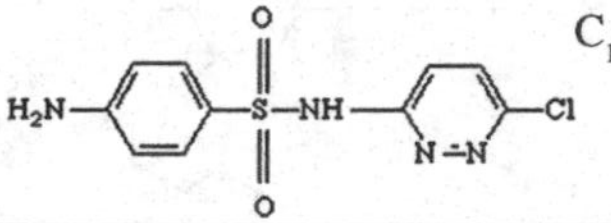	$C_{10}H_9ClN_4O_2S$	284.74	—	—

β-lactams. The chemical structure of these substances is based on the β-lactam ring. This group includes penicillin derivatives, β-lactamase inhibitors, and cephalosporins and other subfamilies such as cephamycines and clavulanic acid (Table 1.8). They act by disrupting the growth of gram-positive bacteria by disrupting the development of bacterial cell walls. The β-lactams can also increase feed efficiency and thus promote growth.

Tetracyclines. These are broad-spectrum antibiotics with high activity against gram-positive and gram-negative bacteria, derived from certain *Streptomyces* spp., that act on bacterial protein synthesis. They can be used to treat respiratory diseases in farm animals. At low doses they can promote growth in animals. Tetracycline, oxytetracycline, and chlortetracycline are some of the most well-known compounds in this group used in veterinary medicine (Table 1.8).

Aminoglycosides. These antibiotics, which have a broad spectrum of activity, act against the synthesis of bacterial cell proteins in gram-negative bacteria. They are based on aminosugars linked by glycoside bridges to a central aglycone moiety. Streptomycin and dihydrostreptomycin belong to the streptomycin subgroup, whereas gentamicin and neomycin belong to the deoxystreptamine subgroup (Table 1.9). They have different subclasses depending on the substituents of the deoxystreptamine moiety (i.e., neomycin A, B, or C).

Macrolides. These act against gram-positive bacteria and are used to treat respiratory diseases. Their structure is based on a macrocyclic lactone ring having carbohydrates attached. They are produced from certain *Streptomyces* strains. Erythromycin is a good representative of this group. Tylosin, spiramycin, and lincomycin are also typical compounds belonging to this group that have been used for growth promotion (Table 1.10).

Quinolones. These act against the bacterial DNA-gyrase with a broad antibacterial activity. Oxolinic acid, flumequine, and nalidixic acid are compounds of the first generation. They are synthesized from 3-quinolone carboxylic acid. The second-generation compounds, which are more potent, are fluoroquinolones like sarafloxacin, enrofloxacin, and danofloxacin, which display fluorescence (Table 1.11). These substances are poorly soluble in water at neutral pH but increase their solubility at basic pH.

Peptides. These are large and complex molecules that are obtained from bacteria and molds. They include nisin, bacitracin, colistin, avoparcin, polymirxin, and Virgiamycin (Croubels et al. 2004). They interact with the bacterial cell wall, resulting in cell membrane damage. These antibiotics often have a mixture of several molecules (i.e., bacitracin A or F). Avoparcin was banned in the EU in 1997, and bacitracin and virgiamycin were banned in 1999 due to the risk of transmission of antibiotic resistance to bacteria (Verdon 2008).

Amphenicols. These are broad-spectrum antibiotics. Chloramphenicol, thiamphenicol, and fluorphenicol are the main representatives of this group. Chloramphenicol was banned in the late 1980s due to its toxic effects.

Table 1.8 Main properties of relevant β-lactam antibiotics and tetracyclines (Reig and Toldrá 2009a)

Substance	IUPAC name	CAS number	Structure	Formula	Molecular mass (g/mol)	Solubility in water (g/mL)
Lactam antibiotics						
Amoxicillin	[2 S-[2α,5α,6α(S*)]]-6-[[amino(4-hydroxyphenyl) acetyl]amino]-3,3-dimethyl-7-oxo-4-thia-1-azabicyclo[3.2.0] heptane-2-carboxylic acid	26787-78-0	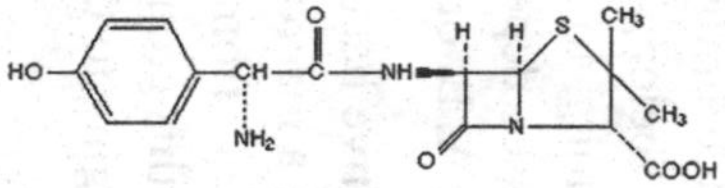	$C_{16}H_{19}N_3O_5S$	365.41	Slightly soluble
Penicillin G calcium	[2 S-(2α,5α,6α)]-3,3-dimethyl-7-oxo-6[(phenylacetyl) amino]-4-thia-1-1-azabicyclo-[3.2.0] heptane-2-carboxylic acid calcium salt	61-33-6	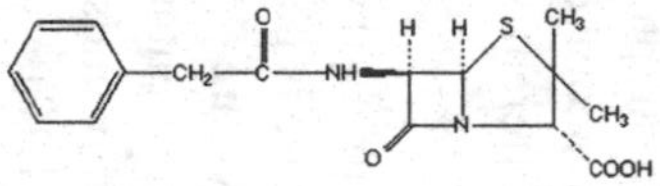	$(C_{16}H_{17}N_2O_4S)_2Ca$	706.84	Soluble
Penicillin V	3,3-dimethyl-7-oxo-6-[(phenoxyacetyl) amino]-4-thia-1-azabicyclo[3.2.0] heptane-2-carboxylic acid	87-08-1	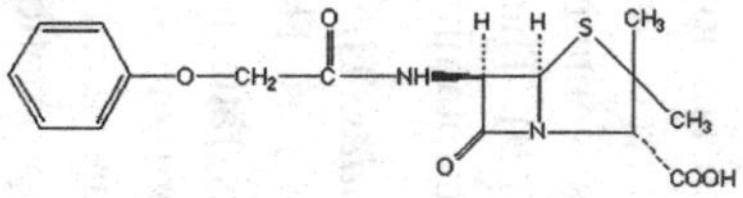	$C_{16}H_{18}N_2O_5S$	350.38	Slightly soluble in acid water

Tetracyclines						
Tetracycline	4-(dimethylamino)-1,4,4a,5,5a,6,11,12a-octahydro-3,6,10,12,12a-pentahydroxy-6-methyl-1,11-dioxo-2-naphthacenecarboxamide	60-54-8	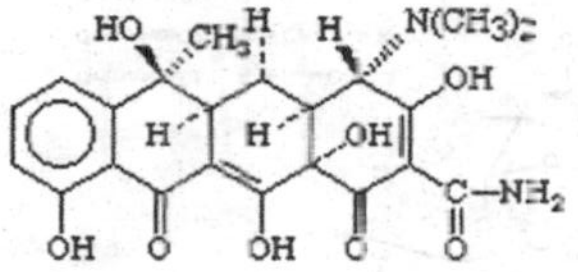	$C_{22}H_{24}N_2O_8$	444.43	—
Oxytetracycline	4-(dimethylamino)-1,4,4a,5,5a,6,11,12a-octahydro-3,5,6,10,12,12a-hexahydroxy-6-methyl-1,11-dioxo-2-naphthacene-carboxamide	79-57-2	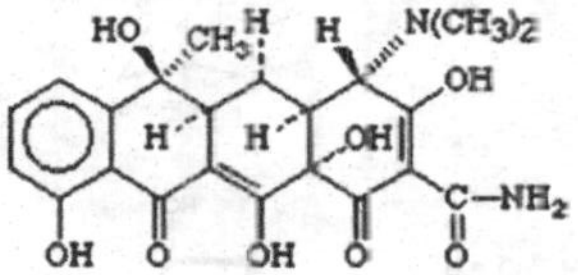	$C_{22}H_{24}N_2O_9$	460.44	—
Chlortetracycline	7-chloro-4-dimethylamino-1,4,4a,5,5a,6,11,12a-octahydro-3,6,10,12,12a-pentahydroxy-6-methyl-1,11-dioxo-2-naphthacene-carboxamide	57-62-5	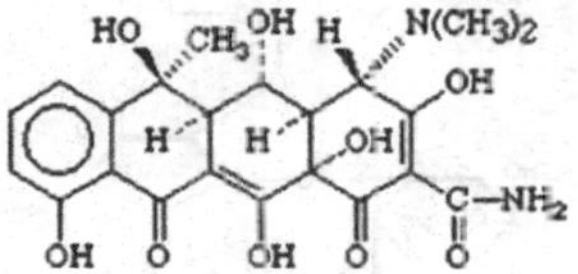	$C_{22}H_{23}ClN_2O_8$	478.88	Slightly soluble

Table 1.9 Main properties of relevant aminoglycosides (Reig and Toldrá 2009a)

Substance	IUPAC name	CAS number	Structure	Formula	Molecular mass (g/mol)	Solubility in water (g/mL)
Dihydros-treptomycin	O-2-deoxy-2-(methylamino)-α-L-glucopyranosyl-(1Ø2)-O-5-deoxy-3-C-(hydroxymethyl)-α-L-lyxofuranosyl-(1Ø4)-N, N'-bis(aminoiminomethyl)-D-streptamine	128-46-1	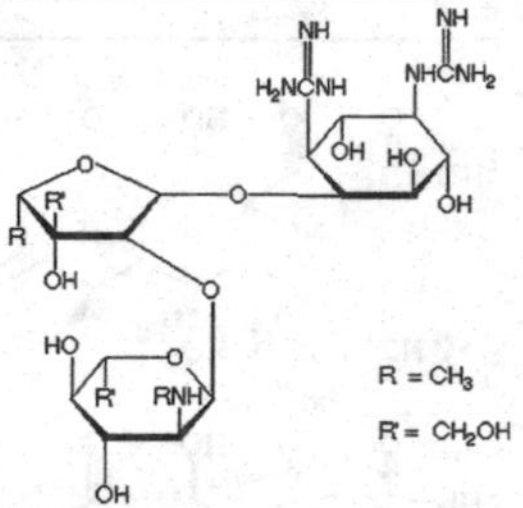 $R = CH_3$; $R' = CH_2OH$	$C_{21}H_{41}N_7O_{12}$	583.62	Soluble
Gentamycin	Various: gentamycin C_1; gentamycin C_2; gentamycin C_{1a} or D; gentamycin A	1403-66-3	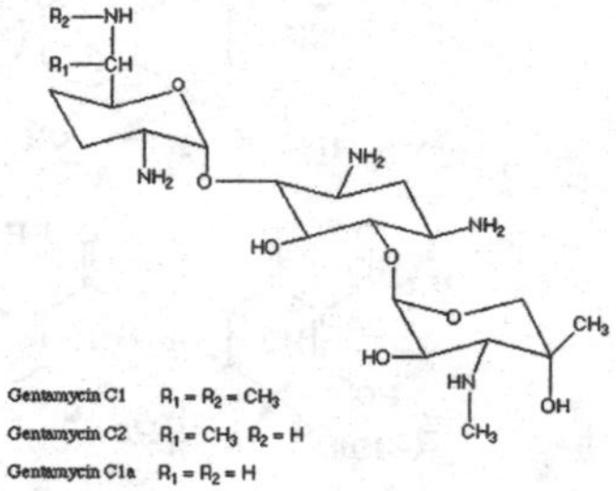 Gentamycin C1 $R_1 = R_2 = CH_3$; Gentamycin C2 $R_1 = CH_3$ $R_2 = H$; Gentamycin C1a $R_1 = R_2 = H$	Several	Several	Soluble

Streptomycin	O-2-deoxy-2-(methylamino)-α-L-glucopyranosyl-(1Ø2)-O-5-deoxy-3-C-formyl-αØ-L-luxofuranosyl-(1Ø4)-N, N'-bis(aminoiminomethyl)-D-streptamine	57-92-1	$R_1 = CH_3$ $R_2 = CH_2OH$ $R_3 = CH_3NH$ $R_4 = CHO$	$C_{21}H_{39}N_7O_{12}$	581.58	Soluble
Streptomycin B		128-45-0	$R_1 = CH_3$ $R_2 = CH_2OH$ $R_3 = CH_3NH$ $R_4 = CHO$	$C_{27}H_{49}N_7O_{17}$	743.72	Soluble

Table 1.10 Main properties of macrolides (Reig and Toldrá 2009a)

Substance	IUPAC name	CAS number	Structure	Formula	Molecular mass (g/mol)	Solubility in water (g/mL)
Tylosin	2-[12-[5-(4,5-dihydroxy-4, 6-dimethyl-oxan-2-yl)oxy-4-dimethylamino-3-hydroxy-6-methyl-oxan-2-yl]oxy-2-ethyl-14-hydroxy -3-[(5-hydroxy-3,4-dimethoxy-6-methyl-oxan-2-yl)oxymethyl]-5,9,13-trimethyl-8,16-dioxo-1-oxacyclohexadeca-4,6-dien-11-yl]acetaldehyde	1401-69-0	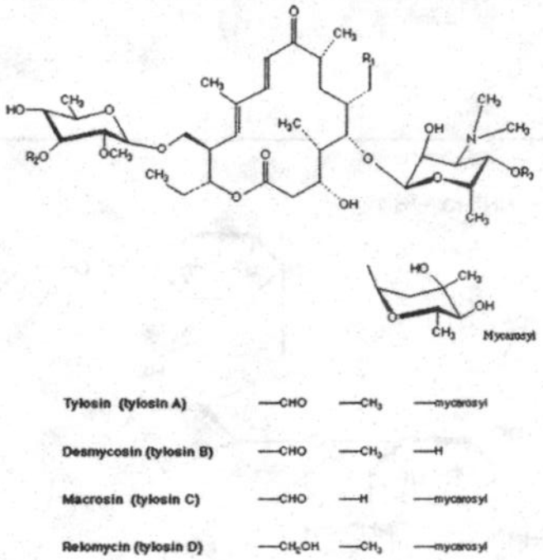 Tylosin (tylosin A) —CHO —CH3 —mycarosyl Desmycosin (tylosin B) —CHO —CH3 —H Macrosin (tylosin C) —CHO —H —mycarosyl Relomycin (tylosin D) —CH2OH —CH3 —mycarosyl	$C_{46}H_{77}NO_{17}$	916.14	Soluble
Erythromycin	(2R,3 S,4 S,5R,6R,8R,10R,11R, 12 S,13R)-5-(3-amino-3,4, 6-trideoxy-N,N-dimethyl-β-D-xylo-hexopyranosyloxy)-3-(2,6-dideoxy-3-C,3-O-dimethyl-α-L-ribo-hexopyranosyloxy)-13-ethyl-6,11,12-trihydroxy-2,4,6,8, 10,12-hexamethyl-9-oxotridecan-13-olide	114-07-8	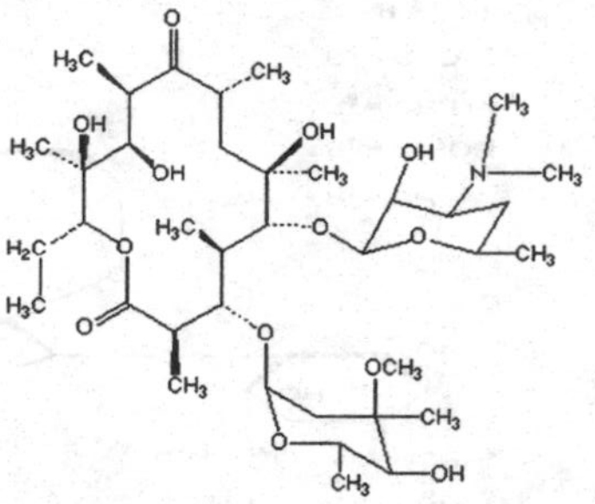	$C_{37}H_{67}NO_{13}$	733.92	Fairly soluble (0.25)

Spiramycin	Complex: spiramycin I or foromacidin A; spiramycin II or foromacidin B; spiramycin III or foromacidin C	8025-81-8	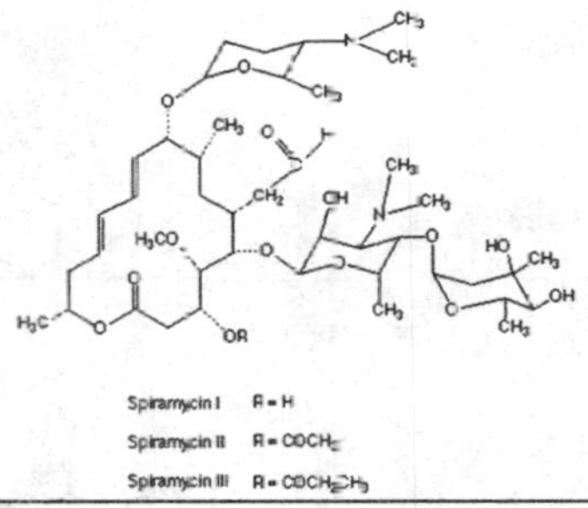 Spiramycin I R = H Spiramycin II R = $COCH_3$ Spiramycin III R = $COCH_2CH_3$	$C_{43}H_{74}N_2O_{14}$	843.05	Slightly soluble

Table 1.11 Main properties of relevant quinolones, and synthetic quinoloxoline compounds (Reig and Toldrá 2009a)

Substance	IUPAC name	CAS number	Structure	Formula	Molecular mass (g/mol)	Solubility in water (g/mL)
Quinolones						
Enrofloxacin	1-cyclopropyl-7-(4-ethyl-1-piperazinyl)-6-fluoro-1,4-dihydro-4-oxo-3-quinolinecarboxilic acid	93106-60-6	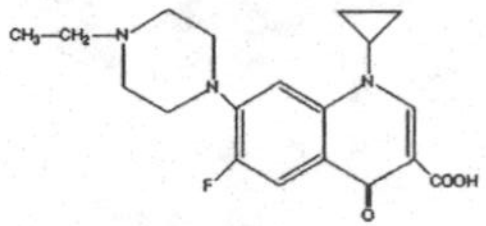	$C_{19}H_{22}FN_3O_3$	359.40	Slightly soluble
Sarafloxacin	6-fluoro-1-(4-fluorophenyl)-1,4-dihydro-4-oxo-7-(1-piperazinyl)-3-quinolinecarboxylic acid	98105-99-8	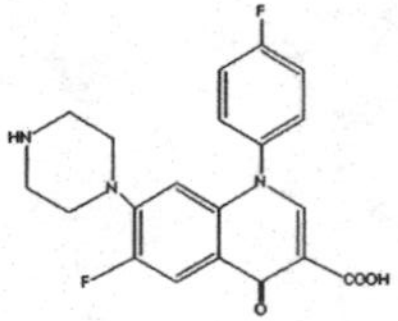	$C_{20}H_{17}F_2N_3O_3$	385.37	Fairly soluble
Danofloxacin	(1 S)-1-cyclopropyl-6-fluoro-1,4-dihydro-7-(5-methyl-2,5-diazabicyclo[2.2.1]hept-2-yl)-4-oxo-3-quinolinecarboxylic acid	112398-08-0	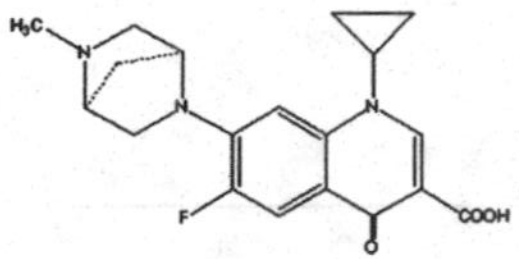	$C_{19}H_{20}FN_3O_3$	357.38	Fairly soluble
Quinoxolines						
Carbadox	(2-quinoxalinylmethylene) hydrazinecarboxylic acid methyl ester N,N'-dioxide; 3-(2-quinoxalinylmethylene) carbazic acid methyl ester N,N'-dioxide	6804-07-5	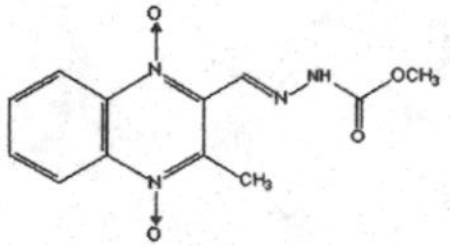	$C_{11}H_{10}N_4O_4$	262.22	Insoluble

Olaquindox	N-(2-hydroxyethyl)-3-methyl-2-quinoxaline-carboxamide 1,4-dioxide	23696-28-8		$C_{12}H_{13}N_3O_4$	263.25	Slightly soluble
Cyadox	[(1,4-Dioxido-2-quinoxalinyl) methylene]hydrazide cyanoacetic acid	65884-46-0		$C_{12}H_9N_5O_3$	271.23	—

Carbadox, olaquindox, and cyadox. These are antibacterial synthetic quinoxaline compounds that have been used as growth promoters (Table 1.11). Carbadox has shown mutagenic and carcinogenic effects in animals, and olaquindox is strongly mutagenic (Croubels et al. 2004). Both antibiotics are rapidly converted into quinoxaline-2-carboxylic acid (QCA) and methyl-3-quinoxaline-2-carboxylic acid (MQCA), respectively. These metabolites are mutagenic and carcinogenic (Verdon 2008). Cyadox is a quinoxaline-N-dioxide that promote growth in poultry and promote feed conversion (Huang et al. 2008). It has been reported that it shows little toxicity but is metabolized in pigs and goat into its desoxy derivatives like 4-desoxycyadox, 1,4-bisdesoxycyadox, cyadox-1-monoxide, and cyadox-4-monoxide and into carboxylic acid derivatives that are further metabolized into quinoxaline-2-carboxylic acid (Zhang et al. 2005; He et al. 2011), which, as was mentioned previously, is mutagenic and carcinogenic (Verdon 2008).

Nitrofurans. These are synthetic compounds with a broad spectrum of activity against bacteria. The main representative substances are furazolidone, furaltadone, nitrofurazone, and nitrofurantoin (Table 1.12). These substances are used against gastrointestinal infections in farm animals but were banned due to their genotoxic and mutagenic properties. They are rapidly metabolized in the organism (i.e., semicarbazide from nitrofurazone), making its detection more difficult.

1.3.4 Other Veterinary Drugs

Antihelmintic agents. The feces of animals may contain eggs or larvae from worm parasites (helminths) that can be ingested by other animals, especially cattle and sheep, with pasture. These drugs act on the metabolism of the parasite. Several groups such as benzimidazoles (thiabendazole, albendazole) imidazothiazoles (tetramisole, levamisole), avermectins (ivermectin, doramectin), and anilides (oxyclozanide, rafoxanide, and nitroxynil) once had widespread use.

Anticoccidials, including nitroimidazoles. Coccidia parasites are transmitted by fecal infection, especially on farms. Anticoccidials are used in poultry to prevent and control coccidiosis, a contagious infection carried by parasites that causes serious effects such as bloody diarrhea and loss of egg production. There are several groups of anticoccidiosis compounds such as nitrofurans, carbanilides, 4-hydroxyquinolones, pyrimidines, and ionophores. Ionophores are polyether antibiotics used against coccidia parasites in poultry. They include monensin, salinomycin, narasin, and lasalocid.

Nitroimidazoles are obtained synthetically with a structure based on a 5-nitroimidazole ring. Main compounds are dimetridazole, metronidazole, ronidazole, and ipronidazole. They are toxic to bacteria when the 5-nitro group is reduced to free radicals by the nitro reductase of anaerobic bacteria (Verdon 2008). These compounds are mutagenic, carcinogenic, and toxic to eukaryotic cells and, thus, have been banned in the EU since the 1990s for use in food-producing animals.

Table 1.12 Main properties of relevant nitrofurans (Reig and Toldrá 2009a)

Substance	IUPAC name	CAS number	Structure	Formula	Molecular mass (g/mol)	Solubility in water (g/mL)
Furaltadone	5-(4-morpholinylmethyl)-3-[[(5-nitro-2-furanyl) methylene]amino]-2-oxazolidinone	139-91-3	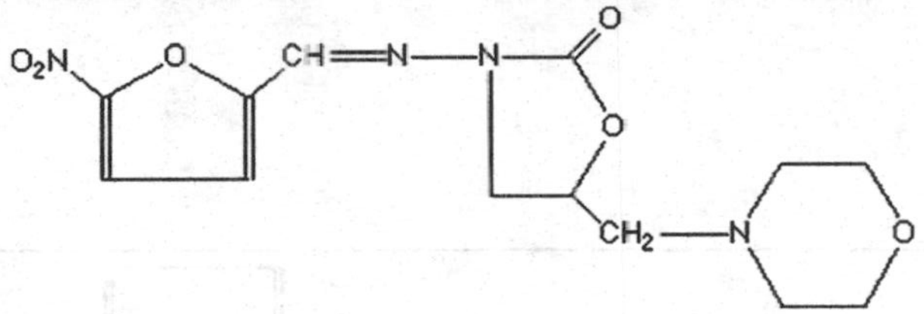 O_2N … CH=N—N … O … CH_2—N … O	$C_{13}H_{16}N_4O_6$	324.29	Slightly soluble
Furazolidone	3-[[(5-nitro-2-furanyl) methylene]-amino]-2-oxazolidinone	67-45-8	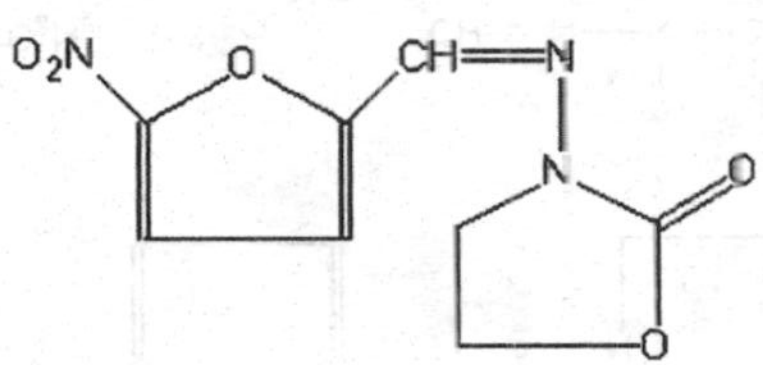 O_2N … O … CH=N … N … O	$C_8H_7N_3O_5$	225.16	Very slightly soluble

(continued)

Table 1.12 (continued)

Substance	IUPAC name	CAS number	Structure	Formula	Molecular mass (g/mol)	Solubility in water (g/mL)
Nitrofurantoin	1-[[(5-nitro-2-furanyl) methylene]amino]-2,4-imidazolidinedione	67-20-9	O_2N, O, CH=N, N, O, NH, O	$C_8H_6N_4O_5$	238.16	Very slightly soluble
Nitrofurazone	2-[(5-nitro-2-furanyl) methylene]-hydrazinecarboxamide	59-87-0	O_2N, O, CH=N–N(H)–C(=O)–NH_2	$C_6H_6N_4O_4$	198.14	Very slightly soluble

Sedatives. These compounds are used to regulate stress in farm animals, but after several weeks they can also induce growth by the redistribution of fat to muscle tissue. Compounds include carazolol, chlorpromazine, azarperone, and xylazine.

1.3.5 Control of Residues of Growth Promoters and Antibiotics in Meat and Poultry

The detection of residues of veterinary drugs is a complex task because of the large number of substances to be assayed, the large number of samples to be analyzed, usually in a restricted period of time, and the low levels of the substances to be detected.

In the USA, the National Residue Program (NRP), administered by the US Department of Agriculture (USDA) Food Safety and Inspection Service (FSIS), oversees the control of veterinary drug residues in the USA under two programs. (1) The FSIS domestic residue sampling program is focused on preventing the occurrence of violative residues in food-producing animals; thus, several sampling plans are in place to verify and ensure that slaughter establishments are fulfilling their responsibilities under the Hazard Analysis and Critical Control Points regulation and according to the regulations of the Food and Drug Administration (FDA) and the Environmental Protection Agency (EPA). (2) The FSIS import residue testing program is focused on determining the effectiveness of exporting countries' residue control programs. FSIS also establishes the type of protocols, and inspector generated in-plant residue test samples (Croubels et al. 2004). The FDA Center for Veterinary Medicine issues the analytical criteria. The control of residues of these substances in meats exported to the EU was further assured by an additional testing program designed by the USDA (Croubels et al. 2004). Part Number 556 under title 21, Food and Drugs of the Code of Federal Regulations, gives the tolerances for residues of new animal drugs in foods (National Archives and Records Administration 2008). The tolerances are based on residues of drugs in edible products of food-producing animals treated with such drugs (Byrnes 2005).

Some growth-promoting substances like estradiol, progesterone, and testosterone are allowed in the USA and other countries like Canada, Mexico, Australia, and New Zealand but under strict application measures and acceptable withdrawal periods. On the other hand, the use of growth promoters has been officially banned in the EU since 1988 (European Community 1988), and only some of them can be permitted for specific therapeutic purposes under strict control and administration by a veterinary officer (Van Peteguem and Daeselaire 2004). In the EU, the monitoring of residues of substances having hormonal or thyreostatic action as well as β-agonists is regulated through Council Directive 96/23/EC (European Community 1996). Member states have set up national monitoring programs and sampling procedures following this directive, which establishes the measures for monitoring certain substances and residues in live farm animals and derived animal products. The main veterinary drugs and substances with anabolic effects as defined in such

Table 1.13 Veterinary drugs and some representative substances with anabolic effect according to European Union classification (EC 1996)

Group A: Substances having anabolic effect	Representative substances
1 Stilbenes	Diethylstilbestrol
2 Anthithyroid agents	Thiouracils, mercaptobenzimidazoles
3 Steroids	
Androgens	Trenbolone acetate
Gestagens	Melengestrol acetate
Estrogens	17-β-estradiol
4 Resorcycilic acid lactones	Zeranol
5 β-agonists	Clenbuterol, mabuterol, salbutamol
6 Other substances	Nitrofurans
Group B: Veterinary drugs	
1 Antibacterial substances	Sulfonamides, tetracyclines, β-lactam, macrolides (tylosin), quinolones, aminoglycosides, carbadox, olaquindox
2 Other veterinary drugs	
Antihelmintics	Benzimidazoles, robenzimidazoles, piperazines, imidazothiazoles, avermectins, etrahydropyrimidines, anilides
Anticoccidials	Nitroimidazoles, carbanilides, 4-hydroxyquinolones, pyridinols, ionophores
Carbamates and pyrethroids	Esters of carbamyc acid, type 1 and 2 pyrethroids
Sedatives	Butyrophenones, promazines, β-blocker carazolol
Nonsteroideal anti-inflammatory drugs	Salicylates, pyrazolones, nicotinic acids, phenamates, arylpropionic acids, pyrrolizines
Other pharmacologically active substances	Dexamethasone
Group B: Contaminants	
3 Environmental contaminants	
Organochlorine compounds	PCBs, compounds derived from aromatic, ciclodiene or terpenic hydrocarbons
Organophosphorous compounds	Malathion, phorate
Chemical elements	Heavy metals
Mycotoxins	Aflatoxins, deoxynivalenol, zearalenone
Dyes	
Others	

directives are given in Table 1.13. Group A includes unauthorized substances with anabolic effects, whereas group B includes veterinary drugs, some of which have established MRLs. The MRL is based on the type and amount of residual substance in the foodstuff that constitutes no risk for consumers (European Community, 2001). MRLs may differ from one international authority to another. As some substances are metabolized in the organism into certain metabolites that can be good markers, the monitoring of residues should serve as a control of the active substance, its degradation products, and its metabolites that may remain in the foodstuffs (Bergwerff and Schloesser 2003; Bergwerff 2005; Toldrá and Reig, 2012).

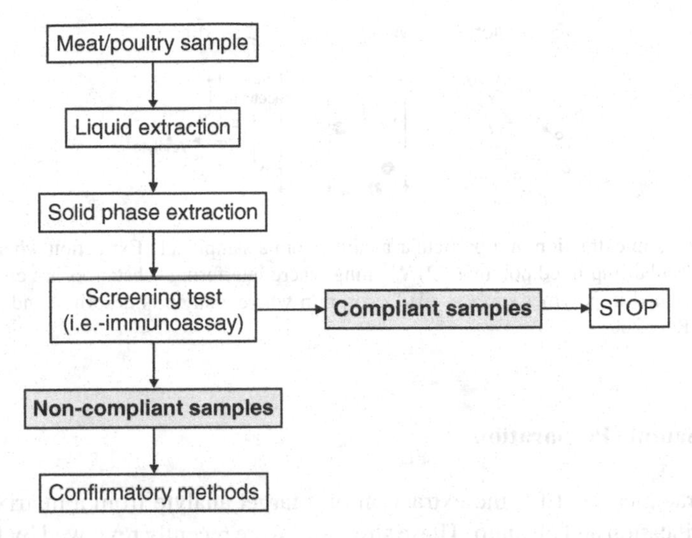

Fig. 1.1 Example of a typical standard procedure for routine control of residues in meat samples (Adapted from Reig and Toldrá (2008a))

1.3.6 Analytical Methodologies for Detection of Veterinary Drugs

In the EU, Commission Decisions 93/256/EC (European Community 1993a) and 93/257/EC (European Community 1993b) established the criteria that the analytical methodology should follow for the adequate screening, identification, and confirmation of banned residues. Commission Decision 2002/657/EC (European Community 2002a) implemented Council Directive 96/23/EC (European Community 1996) and has been in force since 1 September 2004. This decision provides rules for the analytical methods to be used in testing official samples and lays down specific criteria by which official control laboratories are to interpret the analytical results of such samples. In the case of screening methods, the correct validation procedures are also stated. An example of a general procedure for the analysis of a meat or poultry sample when screening for veterinary drug residues is shown in Fig. 1.1.

These regulations usually imply the analysis of very different types of residues (e.g., agonists, thyreostatic agents, various antibiotics) in a variety of matrices such as feed, water, urine, hair, muscle, and organs and in a large number of samples, necessitating the availability of screening techniques (Bergwerff 2005; Reig and Toldrá 2008a). The initial control is usually based on screening tests like ELISA test kits, lateral flow sticks, antibody-based automatic techniques, or chromatographic techniques (Toldrá and Reig 2006; Reig and Toldrá 2009a; Cháfer et al. 2010). Screening tests are rapid and, in the case of immunoassays, have a high specificity and sensitivity, but unfortunately, they are unable to confirm results because they can only yield qualitative or semiquantitative data (Reig and Toldrá 2008b).

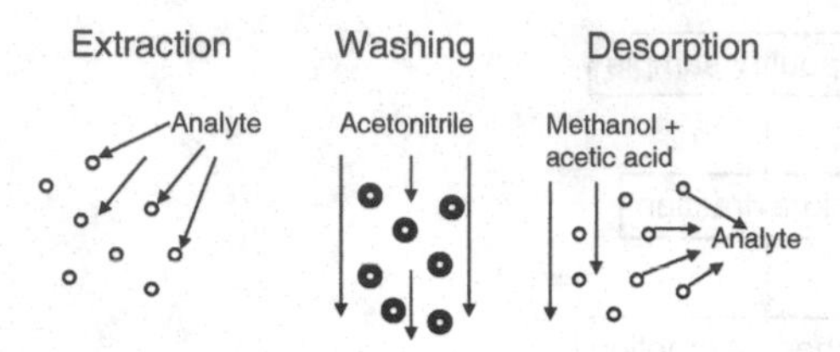

Fig. 1.2 Stages in extraction of a particular analyte from a sample. (1) Extraction where analyte binds to molecular imprinted polymer. (2) Washing where interfering substances are eluted while analyte is retained in polymer surface. (3) Desorption where analyte is desorbed and recovered (Toldrá and Reig 2008)

1.3.6.1 Sample Preparation

Several strategies exist for the extraction of a target analyte from a matrix and its partial purification and cleanup. These strategies were recently reviewed by Kinsella et al. (2009). Some of these techniques are briefly described below.

Solid-phase extraction. The common cleanup procedures for complex matrices like meat or poultry are based on solid-phase extraction (SPE) techniques that are very fast and economical but have insufficient selectivity. Analytes are extracted by partitioning between a solid sorbent surface and the liquid phase (sample). Polymeric SPE cartridges are usually used, and automated SPE systems are available. The choice of SPE technique depends on the type of analyte and matrix, which determine the maximum recovery and improve the sensitivity of the analytical method.

Molecularly imprinted solid-phase extraction (MISPE). Several methods based on molecular recognition mechanisms for the cleanup of samples have been developed in recent years (Widstrand et al. 2004; Baggiani et al. 2007). A typical extraction procedure is shown in Fig. 1.2. Molecularly imprinted polymers (MIPs) consist of cross-linked polymers prepared in the presence of a template molecule that can be a specific analyte; such polymers are useful for the isolation of small amounts of residues in meat. MIPs can support high temperatures, wide pH ranges, and a variety of organic solvents. The extracted residues are then analyzed by liquid chromatography-mass spectrometry and have shown good quantitative results for chloramphenicol (Boyd et al. 2007), β-agonists in pork and liver (Hu et al. 2011), cimaterol, ractopamine, clenproperol, clenbuterol, brombuterol, mabuterol, mapenterol, and isoxsurine but not for salbutamol and terbutaline (Berggren et al. 2000; Stubbings et al. 2005; Kootstra et al. 2005).

Immunoaffinity chromatography. This type of chromatography is based on antigen–antibody interactions, which are very specific and valid for the purification of a particular analyte. A scheme of the procedure is shown in Fig. 1.3. The columns are packaged with a solid matrix where a specific antibody for the target analyte is bound. Once the extract is injected into the column, the analyte is retained by the antibody bound to the matrix while the rest of the extract is eluted. The target analyte

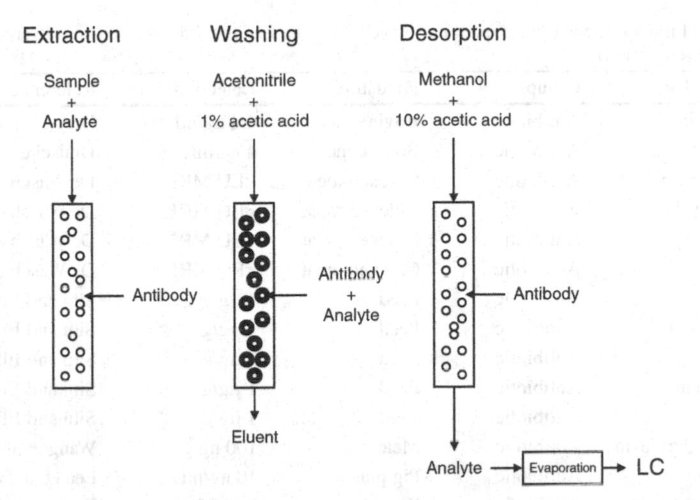

Fig. 1.3 Stages in immunoaffinity chromatography purification of a particular analyte. (**a**) Extraction where analyte binds to antibody immobilized to packaging. (**b**) Washing where interfering substances are eluted while analyte is retained in packaging. (**c**) Desorption where analyte is free from its bound to the antibody and recovered (Toldrá and Reig 2008)

is eluted by an antibody–antigen dissociating buffer and recovered in high concentration (Fig. 1.3). This technique has yielded good results for different residues like zearalenone in feed (Campbell and Armstrong 2007), zeranol (Zhang et al. 2006a), and avermectin (He et al. 2005). Immunoaffinity chromatographic columns can only be reused a limited number of times; furthermore, they are sometimes limited by interference due to cross reactions by other residues of the sample with the antibody (Godfrey 1998).

1.3.6.2 Screening Techniques

Several techniques exist for the screening of residues.

Immunoassay kits. These kits are simple to use and manipulate and are also very specific for a given residue because they are based on antigen–antibody interactions. A decent number of immunoassays have been developed in recent years and are commercially available for the detection of veterinary drug residues in foods. These methods are based on enzyme-linked immunosorbent assays (ELISA), enzyme immunoassay (EIA), lateral flow immunoassays, radio immunoassay (RIA), and arrays and chips (biosensors). With ELISA or EIA kits, detection is based on a change in color that is proportional to the amount of target analyte present in the sample. A similar change in color is the basis for dipsticks, which consist of an antibody immobilized at the end of a plastic stick (Link et al. 2007; Levieux 2007). The use of luminescence or fluorescence detectors may increase the sensitivity (Roda et al. 2003; Zhang et al. 2006b). The limits of detection depend on the

Table 1.14 Limits of detection or quantitation (CCβ) of ELISA test kits assayed for different residues (Toldrá and Reig 2008)

Type of residue	Group	Foodstuff	Detection limit[a]	Reference
Erythromycin	Antibiotic	Bovine meat	0.4 ng/mL	Draisci et al. 2001
Tylosin	Antibiotic	Bovine meat	4 ng/mL	Draisci et al. 2001
Oxytetracycline	Antibiotic	Chicken meat	<EU MRL	De Wasch et al. 2001
Chlortetracycline	Antibiotic	Chicken meat	<EU MRL	De Wasch et al. 2001
Doxyckine	Antibiotic	Chicken meat	<EU MRL	De Wasch et al. 2001
Tetracycline	Antibiotic	Chicken meat	<EU MRL	De Wasch et al. 2001
Bacitracin	Antibiotic	Feed	1 μg/g	Situ and Elliott 2005
Tylosin	Antibiotic	Feed	1 μg/g	Situ and Elliott 2005
Spiramycin	Antibiotic	Feed	1 μg/g	Situ and Elliott 2005
Virginiamycin	Antibiotic	Feed	1 μg/g	Situ and Elliott 2005
Olaquindox	Antibiotic	Feed	1 μg/g	Situ and Elliott 2005
Sulphachlorpyridazine	Antibiotic	Meat	100 ng/g	Wang et al. 2006
Tetracycline	Antibiotic	Pig plasma	10 ng/mL	Lee et al. 2001
Tylosine	Antibiotic	Water	0.1 ng/mL	Kumar et al. 2004
Tetracycline	Antibiotic	Water	0.05 ng/mL	Kumar et al. 2004
Chloramphenicol	Antibiotic	Chicken muscle	6 ng/L	Zhang et al. 2006a
Diethylestilbestrol	Estrogen	Chicken meat	0.07 ng/mL	Xu et al. 2006a
Hexoestrol	Estrogen	Pork meat	0.07 ng/mL	Xu et al. 2006b
Avermectins	Insecticidal	Bovine liver	1.06 ng/mL	Shi et al. 2006
Medroxyprogesterone acetate	Steroid	Meat	0.096 ng/g	Chifang et al. 2006
Semicarbazide	Nitrofuran	Chicken meat	CCβ = 0.25 ng/g	Cooper et al. 2007a
Dimetridazole	Nitroimidazoles	Chicken muscle	CCβ = 2 ng/g	Huet et al. 2005
Metronidazole	Nitroimidazoles	Chicken muscle	CCβ = 10 ng/g	Huet et al. 2005
Ronidazole	Nitroimidazoles	Chicken muscle	CCβ = 20 ng/g	Huet et al. 2005
Hydroxydimetridazole	Nitroimidazoles	Chicken muscle	CCβ = 20 ng/g	Huet et al. 2005
Ipronidazole	Nitroimidazoles	Chicken muscle	CCβ = 40 ng/g	Huet et al. 2005
Azaperol	Sedative	Pork kidney	5 ng/g	Cooper et al. 2007b
Azaperone	Sedative	Pork kidney	15 ng/g	Cooper et al. 2007b
Carazolol	Sedative	Pork kidney	5 ng/g	Cooper et al. 2007b
Acepromazine	Sedative	Pork kidney	5 ng/g	Cooper et al. 2007b
Chlorpromazine	Sedative	Pork kidney	20 ng/g	Cooper et al. 2007b
Propionylpromazine	Sedative	Pork kidney	5 ng/g	Cooper et al. 2007b

[a] Limits of detection or CCβ

previous extraction and cleanup of the sample (De Wasch et al. 2001; Gaudin et al. 2003; Cooper et al. 2004). Some false positives may arise as a result of interference from other substances present in the sample. In any case, when there is any doubt or uncertainty, samples must be submitted to confirmatory analysis for further confirmation. Several examples of assayed or developed ELISA test kits for the detection of different residues and their respective limits of detection or quantitation are shown in Table 1.14.

Several interlaboratory tests have been performed to check and compare the validity of the different kits from different suppliers and for specific residues, revealing generally good results (Gaudin et al. 2003; Situ et al. 2006; Cooper et al. 2003). However, ELISA test kits cannot be used for multiresidue analysis and have also witnessed large cost increases, making its use somehow restrictive.

Biosensors. These instruments are based on the interaction of an immobilized antibody on the surface of a transducer that interacts with the analyte in the sample (Wang et al. 2006) and then converted into a measurable signal (De Wasch et al. 2001; Draisci et al. 2001). For instance, surface plasmon resonance (SPR) measures variations in the refractive index of a solution adjacent to a metal surface (Cooper et al. 2004; Dumont et al. 2006; Haughey and Baxter 2006). Biosensors have been applied to the rapid detection of veterinary drugs in foods of animal origin. This is a high-throughput technique because it has the capability for simultaneous detection of multiple residues in a sample (Kumar et al. 2004). Biosensors have been used in the detection of various veterinary drug residues like ractopamine (Thompson et al. 2008), nitroimidazoles (Situ and Elliott 2005; Connolly et al. 2007; Cooper et al. 2007a), clenbuterol in urine (Haughey et al. 2001), flumequine in broiler muscle (Haasnoot et al. 2007), chloramphenicol in poultry (Ferguson et al. 2005), chloramphenicol glucuronide in kidney (Ashwin et al. 2005), and sulphonamides in pork (McGrath et al. 2005; Bienemann-Ploum et al. 2005). Other biosensors are based on the use of biochip molecule microarrays that use small molecules as probes immobilized on a variety of surfaces. Detection of clenbuterol, chloramphenicol and tylosin (Peng and Bang-Ce 2006), chloramphenicol (Gaudin et al. 2003), or nitroimidazoles (Huet et al. 2005) has been reported.

Liquid chromatography. High-performance liquid chromatography (HPLC) is seeing expanded use as a screening tool in control laboratories because it allows for the simultaneous analysis of multiple residues in a sample in a relatively short time, especially with the advent of ultra performance liquid chromatography (UPLC). Thus, HPLC has been successfully used for the screening of substances with anabolic properties in different matrices such as quinolone residues in meat and animal tissues (Kirbis et al. 2005; Verdon et al. 2005), sulphonamides in feed (Borràs et al. 2011b), methyl thiouracils (Reig et al. 2005), growth promoters (Koole et al. 1999) and anabolic steroids in urine (Gonzalo-Lumbreras and Izquierdo-Hornillos 2000), and corticosteroids like dexamethasone in water, feed, and meat (Stolker et al. 2000; Reig et al. 2006)

1.3.6.3 Confirmatory Methods

The next step for those suspicious samples (suspected of being noncompliant) consists in a confirmatory analysis through gas chromatography (GC) or HPLC coupled to mass spectrometry or other sophisticated methodologies for accurate identification and confirmation of the substance (Toldrá and Reig 2006). Commission Decision 2002/657/EC (European Community 2002a) implemented Council Directive 96/23/EC (European Community 1996) and has been in force since 1 September 2004. This decision establishes a minimum number of identification points required for the correct

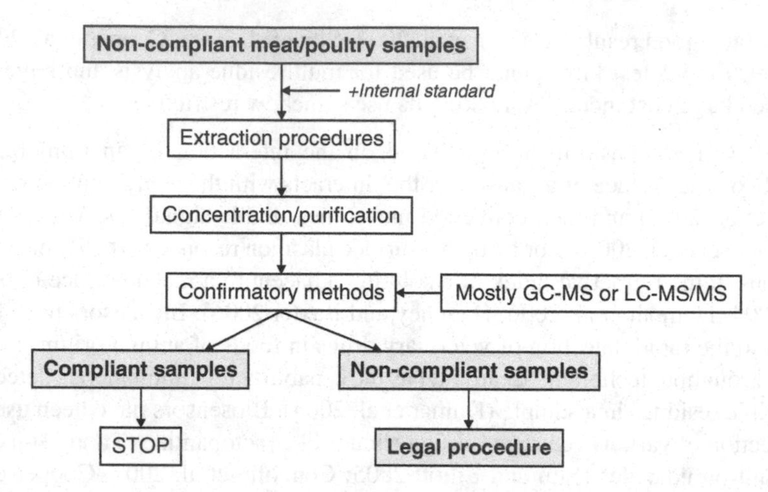

Fig. 1.4 Example of a typical standard procedure for confirmatory analysis of residues in meat or poultry samples found to be noncompliant after screening tests

identification of a substance, and these points are obtained depending on the analytical technique used. For instance, four identification points are earned when using mass spectrometry for the detection of substances in group A and three in the case of substances from group B. In this way, one identification point can be earned for a precursor ion when one uses a triple quadrupole spectrometer and 1.5 points for each product ion; however, if one uses a high-resolution mass spectrometer, two identification points are earned for the precursor ion and 2.5 for each product ion. Another requirements is that the relative retention of the analyte must correspond to that of the calibration solution at a tolerance of ±0.5 % for GC and ±2.5 % for LC. The decision (European Community 2002a) defines the level of confidence in routine analytical results through the decision limit (CCα), defined as the limit at and above which it can be concluded with an error probability of α that a sample is noncompliant, and the detection capability (CCβ), which is defined as the smallest content of the substance that may be detected, identified, or quantified in a sample with an error probability of β.

Confirmatory methods are useful for the identification of a substance so that the sample can be considered as noncompliant (unfit for human consumption) when quantified above the decision limit for a forbidden substance, like those of group A, or exceeding the MRL in the case of substances having an MRL. Internal standards are recommended at the beginning of the extraction procedure (Reig and Toldrá 2011). An example of the procedure to follow for a sample found to be non-compliant after screening tests is shown in Fig. 1.4.

Mass spectrometry coupled to LC is becoming an essential tool for the analysis of residues in meat and poultry, especially for nonvolatile or thermolabile substances. Tandem mass spectrometry (MS-MS) has a high selectivity and sensitivity and thus allows the selection of a precursor m/z (mass to charge ratio), which is performed first. This has several advantages: it eliminates any uncertainty regarding the origin of the observed fragment ions, eliminates any potential interference from the meat matrix or from the mobile phase, and reduces chemical noise (Gentili et al.

2005). Two main types of interfaces can be used depending on the polarity and molecular mass of the analytes: electrospray ionization (ESI) facilitates the analysis of small to relatively large and hydrophobic to hydrophilic molecules (Hewitt et al. 2002; Thevis et al. 2003), whereas atmospheric pressure chemical ionization (APCI) is less sensitive to matrix effects (Puente 2004; Maurer et al. 2004).

Quadrupole time of flight (Q-TOF) has been reported as a useful technique with a better sensitivity and resolution and a high mass accuracy for both precursor and product ions (Van Bocxlaer et al. 2005) that makes it useful for the detection and identification of unknown substances in complex mixtures. The ion suppression phenomenon due to the presence of meat-matrix-interfering compounds that appear to reduce the evaporation efficiency may reduce the analyte detection capability and repeatability (Antignac et al. 2005), leading to the lack of detection of an analyte or the underestimation of its concentration. Correct purification and cleanup of the sample, use of an internal standard, or modification of the elution conditions of the target analyte screening an area not affected by suppression are good preventive measures (Antignac et al. 2005).

Reviews have been published recently about the analysis of antimicrobial substances in animal feed (Borràs et al. 2011a), aminoglycoside and macrolide residues in foods (McGlinchey et al. 2008), growth promoters in meat, poultry, and meat products (Reig and Toldrá 2009b, c), antibiotics in meat (Verdon 2008; Van der Heeft et al. 2009), and veterinary (Le Bizec et al. 2009) and anti-inflammatory drugs in animal foods (Gentili 2007).

Some multiresidue methods for the simultaneous detection of several veterinary drugs and their validation have been recently reported for meat (Kaufmann 2009; Kaufmann et al. 2011) and feed (Cronly et al. 2010); in addition, recent work also includes multiresidue analysis of 16 β-agonists in pig liver, kidney, and muscle (Shao et al. 2009) and the use of hydrophilic interaction liquid chromatograph-tandem mass spectrometry in chicken muscle (Chiaochan et al. 2010). Other authors have reported the analysis of sulphonamides in animal feed by LC with fluorescence detection (Borràs et al. 2011b)

In summary, numerous analytical techniques, including adequate cleanup of samples, are available for the control of the presence of veterinary drug residues, including growth-promoting substances, despite the large variety of matrices (feed, urine, hair, and water on farms and diverse organs and meat in slaughterhouses) where target analytes must be analyzed for correct control. The continuous development of new instrumentation with better sensitivity and other improved capabilities provides adequate tools for the control of such residues at progressively decreasing levels (De Brabander et al. 2009).

1.4 Carcass Disinfectants

Many substances may be used as disinfectants for beef, pork, or poultry carcasses. These substances are quite varied, including chlorine dioxide, acidified sodium chlorite, trisodium phosphate, peroxyacids, or lactic acid (Table 1.15). The efficacy

Table 1.15 Substances used as carcass disinfectants and their properties

Disinfectant	Formula	CAS number	Molecular mass (g/mol)
Chlorine dioxide	ClO_2	10049-04-4	67.45
Acidified sodium chlorite	$NaClO_2$	7758-19-2	91.45
Trisodium phosphate	Na_3PO_4	7601-54-9	163.94
Peroxyoctanoic acid	$C_2H_4O_3$	33734-57-5	75.99
Peroxyacetic acid	$C_8H_{16}O_3$	79-21-0	160.05
Cetylpyridinium chloride	$C_5H_5NC_{16}H_{33}$ Cl	123-03-5	339.99
Lactic acid	$C_3H_6O_3$	50-21-5	90.08

of these antimicrobial substances depends on many factors including the initial microbial load in the carcasses, the concentration of the disinfecting substance, time of exposure, temperature, water pH and hardness, firmness of bacteria attachment to the carcasses, biofilm formation, and the presence of fat or organic material in water. The treatment is able to reduce the contamination level in the carcass but cannot completely eliminate pathogens. The mechanisms of action vary depending on the substance, but in general, microorganisms are killed by action on the cellular membrane and disruption of cellular processes.

The use of chlorine dioxide as a carcass disinfectant, generally at 20–50 ppm, generates chlorite and chlorate as the primary reduction products. Chlorine dioxide concentration decreases rapidly while the concentration of both chlorite and chlorate increases in a 7:3 ratio with increases in the dose of chlorine dioxide and treatment time. Generally, around 5 % of the initial concentration remains as chlorine dioxide (Tsai et al. 1995; United States Department of Agriculture 2002a).

The use of acidified sodium chlorite generates chlorous acid as the primary byproduct but also other substances like chlorite, chlorate, and chlorine dioxide. The proportion depends on the pH of the mixture. Thus, the rate of dissociation of chlorite to chlorous acid is about 31 % at pH 2.3, 10 % at pH 2.9, and 6 % at pH 3.2, and the amount of chlorine dioxide does not exceed 1–3 ppm (USDA 2002b). The initial concentration of sodium chlorite is about 500–1,200 mg/L for spray and dip solutions (pH 2.3–2.9) and 50–150 mg/L for cold water (pH 2.8–3.2).

When trisodium phosphate is used, Na^+ and PO_4^{3-} are the primary ions generated by ionization. The pH of a 1 % solution is 11.5–12.5 (USDA 2002c). The use of lactic acid solution in a concentration of up to 5 % (w/w) has been proposed for the treatment of beef hides.

Peroxyacid solutions consist of a mixture of peroxyacetic acid, peroxyoctanoic acid, hydrogen peroxide, and HEDP (1-hydroxy-1,1-diphosphonic acid) (USDA 2002d). Acetic acid, octanoic acid, water, and oxygen are usually generated when this solution is applied to carcasses, but other compounds such as 1-methoxy-4-methylbenzene, nonanal, and decanal can be generated as well, although in smaller amounts (Monarca et al. 2003, 2004).

Cetylpyridinium chloride is applied in aqueous solution mixed with propylene glycol as a fine mist spray or drench to raw poultry carcasses prior to immersion in a chiller or post chill, at a level not to exceed 0.3 g cetylpyridinium chloride per pound of raw poultry carcass (Li et al. 1997; Food and Agriculture Organization/World Health Organization 2008).

The control of the presence of residues of these disinfecting substances in carcasses following carcass treatment and rinsing is achieved through analytical determinations in carcass samples (meat and fat). Analytical methodologies to detect residues of these substances in carcasses is based on HPLC with UV or diode array detection for the case of water-soluble substances, whereas the analysis of lipid soluble substances is based on GC. When confirmatory analyses are needed, mass spectrometry detectors are coupled to either the HPLC or GC instruments.

1.5 Residues of Environmental Contaminants (Dioxins, Pesticides, Heavy Metals)

Environmental contamination constitutes a huge problem that affects the entire food chain. The main concern for meat and poultry is that such contaminants may be present in the water and feed consumed by farm animals as a route for entering the food chain. There are many types of environmental contaminants. The most relevant are dioxins, organophosphorous and organochlorine pesticides, and heavy metals. Environmental contamination is quite extended worldwide, and globalization makes its control even more difficult. Some of these substances may remain in either animal or human bodies and accumulate, especially in fatty tissues, with long-term effects. Polychlorinated dibenzo-p-dioxins (PCDDs), polychlorinated dibenzofurans (PCDFs), and polychlorinated biphenyls (PCBs), in addition to other related halogenated aromatic compounds, have been identified in the fatty tissues of animals and humans. These substances constitute a group of lipophilic contaminants with low volatility but high stability (Ahlborg et al. 1994).

The United Nations Environment Programme defined the term *persistent organic pollutants* (POPs) to refer to those persistent chemical substances that can accumulate in foods and have adverse effects on human consumers. In fact, some of these contaminants, such as organochlorine pesticides, constitute a real risk of long-term exposure, even though they were banned in the 1970s and 1980s, because they are persistent and stable and remain in the environment for many years (Moats 1994). In the EU, current MRLs for organochloride pesticides in animal products are set within 0.02 and 1 mg/kg of fat (Iamiceli et al. 2009). In the USA, the Environmental Protection Agency (EPA) established tolerances set forth in Title 40 of the Code of Federal Regulations (CFRs). Part 180 establishes the tolerances and exemptions for chemical residues of pesticides in foods (National Archives and Records Administration 2010). Thus, tolerances or exemptions are given for specific categories of food and specific commodities prior to harvest or slaughter meaning each individual food or food group to which the limit applies. This means that it can apply to the parent form of the active ingredient only or to the parent compound with or without one or more metabolites or degradation products or even only the chemical

Table 1.16 TEF values for some dioxins and dioxinlike PCBs (EC 2006a)

Congener	TEF value	Congener	TEF value
Dibenzo-p-dioxin (PCDDs)		*Dioxinlike PCBs: Nonortho PCBs + Mono-ortho PCBs*	
2,3,7,8-TCDD	1		
1,2,3,7,8-PeCDD	1	*Nonortho PCBs*	
1,2,3,4,7,8-HxCDD	0.1	PCB 77	0.0001
1,2,3,6,7,8-HxCDD	0.1	PCB 81	0.0001
1,2,3,7,8-HxCDD	0.1	PCB 126	0.1
1,2,3,4,6,7,8-HpCDD	0.01	PCB 169	0.01
OCDD	0.001		
Dibenzofurans (PCDFs)		*Mono-ortho PCBs*	
2,3,7,8-TCDF	0.1		
1,2,3,7,8-PeCDF	0.05	PCB 105	0.0001
2,3,4,7,8-PeCDF	0.5	PCB 114	0.0005
1,2,3,4,7,8-HxCDF	0.1	PCB 118	0.0001
1,2,3,6,7,8-HxCDF	0.1	PCB 123	0.0001
1,2,3,,7,8,9-HxCDF	0.1	PCB 156	0.0005
2,3,4,6,7,8-HxCDF	0.1	PCB 157	0.0005
1,2,3,4,6,7,8-HpCDF	0.01	PCB 167	0.0001
1,2,3,6,7,8,9-HpCDF	0.01	PCB 189	0.0001
OCDF	0.0001		

T tetra, *Pe* penta, *Hx* hexa, *Hp* hepta, *O* octa, *CDD* chlorobenzodioxin, *CDF* chlorobenzofuran, *CB* chlorobiphenyl

moiety that can be analyzed for calculating the pesticide residue. For instance, in the case of cattle meat, the tolerance is established as 0.1 mg/kg for the carbamate benomyl or 0.05 mg/kg for the organophosphate chloropyrifos (Nielsen 2010).

Feed used for farm animals may contain a large diversity of environmental contaminants like organophosphorous and organochlorine pesticides, dioxins, polychlorinated biphenyls (PCBs), which is a large family (209 compounds) used in lubricating oils and heat exchange fluids, mycotoxins resulting from molds, marine toxins, and heavy metals, among others. The toxic equivalent factor (TEF) was established by the World Health Organization, with the most toxic dioxin having a TEF of 1. The toxic equivalent (TEQ) is obtained through the multiplication of the TEF by the respective PCB concentration (Ahlborg et al. 1994). PCB congeners include nonortho and mono-ortho and are defined as dioxinlike PCBs (Table 1.16). The maximum levels of dioxins in meat and poultry were set in the EU through Council Regulation 1881/2006 (European Commission 2006a). In the case of beef and lamb, such limits are 3.0 pg/g TEQ for total dioxins and 4.5 pg/g TEQ for total dioxins and dioxinlike PCBs. In the case of pork, those limits are 1.0 and 1.5 pg/g, respectively, and 2.0 and 4.0 pg/g, respectively, for poultry (Table 1.17). PCBs may have different effects on humans like dermal toxicity, immunology toxicity, endocrine toxicity, and risk of cancer (Twaroski et al. 2001; Negri et al. 2003; Fenton 2006).

Table 1.17 Maximum levels within EU for environmental contaminants dioxins, dioxinlike PCBs, and heavy metals in several meats and poultry, excluding edible offal (European Community 2005, 2006a)

Substance/food	Bovine	Lamb	Poultry	Pigs
Sum of dioxins (pg TEQ/g fat)	3.0	3.0	2.0	1.0
Sum of dioxins + dioxinlike PBs (pg TEQ/g fat)	3.0	3.0	4.0	1.5
Cadmium (mg/kg w/w)	0.05	0.05	0.05	0.05
Lead (mg/kg w/w)	0.1	0.1	0.1	0.1
Mercury (mg/kg w/w)	0.1	0.1	0.1	0.1

In the case of heavy metals, intake in animals may be via soil and water as well as from feed. Metals of concern are cadmium because of its negative effects on renal and lung as well as cardiovascular and skeletal systems; organic mercury like methylmercury, which can cause brain impairment, anemia, and gastrointestinal complications; arsenic, which can be carcinogenic; and lead, which can damage kidneys and human reproductive and immune systems (Forte and Bocca 2011). The presence of metals in feeding stuffs is regulated in the EU through maximum limits in Directives 2002/32/EC (European Commission 2002b) and 2005/87/EC (European Commission 2005). On the other hand, Regulation 1881/2006 (European Commission 2006a) establishes the limits of metals in foods of animal origin, for instance, less than 0.05 mg Cd/kg and less than 0.1 mg Pb/kg of meat or poultry (Table 1.17).

The reasons for the presence of environmental contamination in meat and poultry are varied: use of contaminated ingredients in feed, lack of control of feed ingredients, inadequate processing, growth of molds in feed grains and meals, etc. (Croubels et al. 2004). The environmental contaminants in meats are difficult to control because of the different potential routes of intake for the animal and the diversity of compounds to be analyzed, even though the contaminants can exert toxicity in the final product (Heggum 2004). Pesticides are generally analyzed with GC or HPLC-based methodologies. The FDA published, and made available on the Internet (FDA 1994), the Pesticide Analytical Manual, which presents the preparation of samples and analytical methodologies for the analysis of pesticides in food. This manual is a repository of the analytical methods used in FDA laboratories to examine food for pesticide residues for regulatory purposes. Volume I contains multiresidue methods routinely used by the FDA because of their efficiency and broad applicability, whereas volume II contains methods designed for the analysis of commodities for residues of only a single compound, usually applied when the likely residue is known. On the other hand, heavy metals are generally analyzed with ICP-MS. The methods for analysis of environmental contaminants in meat, poultry, and derived products are widely reported elsewhere. Recent reviews are available on the methods of analysis for the detection and identification of POPs (Iamiceli et al. 2009), PCBs (García-Regueiro and Castellari 2009), pesticides (Vázquez-Roig and Picó 2011), and heavy metals (Forte and Bocca 2011).

1.6 Substances Generated During Processing of Meat and Poultry

1.6.1 N-Nitrosamines

Nitrosamines are N-nitroso compounds that have attracted much attention in recent decades because of their potential carcinogenic compounds. Nitrosamines are formed in cured meats through the reaction of nitrous acid in its dissociated form (nitrous anhydride) generated from nitrite, with secondary amines. Some of the most important nitrosamines detected in cured meats are N-nitrosodimethylamine, N-nitrosopyrrolidine, N-nitrosopiperidine, N-nitrosodiethylamine, N-nitrosodi-n-propylamine, N-nitrosomorpholine, and N-nitrosoethylmethylamine (Table 1.18). Most of the tested nitrosamines in laboratories are carcinogenic in a wide range of animal species (Rath and Reyes 2009). In addition, a large number of nonvolatile nitroso compounds, higher in molecular weight and more polar, have also been reported. Some of the most important are N-nitrosoamino acids like N-nitrososarcosine and N-nitrosothiazolidine-4-carboxylic acid, hydroxylated N-nitrosamines, N-nitroso sugar amino acids, and N-nitrosamides like N-nitrosoureas, N-nitrosoguanidines, and N-nitrosopeptides (Pegg and Shahidi 2000).

Nitrite is the main additive used as a preservative in cured meats because of its powerful inhibition of the outgrowth of spores of putrefactive and pathogenic bacteria like *Clostridium botulinum*. Nitrite provides other benefits, like its involvement in the generation of nitrosylmyoglobin, which gives the typical pink cured color formation, but also its contribution to the oxidative stability of lipids and indirectly to cured meat flavor (Ramarathnam 1998).

However, the main concern is related to the residual nitrite remaining in the meat product because it can be a source of nitrous acid and thus of nitrosamines if secondary amines are also present (Toldrá et al. 2009). The amount of nitrous acid increases when the pH of the product approaches the pKa of nitrous acid (pKa = 3.36). The amount of N-nitrosamines in meat products depends on many variables like the amount of added and residual nitrite, processing conditions, amount of lean meat in the product, heating if any, and the presence of catalysts or inhibitors (Hotchkiss and Vecchio 1985; Walker 1990). A database with nitrosamine content in 297 food items from 23 countries was recently created with the aim of facilitating the quantification of dietary exposure to potential carcinogens and their relation to certain types of cancer (Jaksyn et al. 2004).

Intense discussions took place in the 1970s about the amounts of residual nitrite remaining in cured meats and the generation of N-nitrosamines in certain cured meat products. It must be taken into account that the generation rate of nitrosamines depends on many variables such as the amount of remaining nitrite, the presence of nitrosation catalysts or inhibitors, the presence of secondary amines, the processing temperature, the pH of the product, time of storage, storage conditions, and the possible addition of reducing substances like ascorbate or isoascorbate (Toldrá and Reig 2007). The presence of microorganisms able to generate nitrite from nitrate via

Table 1.18 List of N-nitrosamines with carcinogenic properties

N-nitrosamines	Structure	CAS number	Molecular mass (g/mol)
N-nitrosodiethylamine		55-18-5	102.14
N-nitrosodiethanolamine		1116-54-7	134.13
N-nitrosopiperidine		100-75-4	114.15
N-nitrosodi-n-propylamine		621-64-7	130.19
N-nitrosodi-n-butylamine		924-16-3	158.24
N-nitrosomethylbenzylamine		937-40-6	150.18

nitrate reductase activity or able to produce amines can also contribute to the generation of nitrosamines. In any case, nitrite is very reactive and rapidly decreases during processing, thus remaining at low residual levels in the final product if correctly processed (Hill et al. 1973). To assure the absence of nitrosamines, it was recommended to reduce the levels of nitrites and add ascorbate or erythorbate to favor the reduction of nitrite to nitric oxide and, thus, the inhibition of nitrosamine formation (Cassens 1997). Ascorbate is better than ascorbic acid because it reacts with nitrite 240 times faster (Pegg and Shahidi 2000). As an example, the residual nitrite content in fermented sausages was found to be below 20 mg/kg in most of the products surveyed in the late 1990s and early 2000s in Europe (European Food Safety Authority 2003). Nitrosodimethylamine and nitrosopiperidine were reported as the main nitrosamines found at levels above 1 μg/kg. The maximum permitted levels for cured meats in the USA are 10 μg/kg (Rath and Reyes 2009). Nitrosamines were assayed in several northern and Mediterranean European fermented sausages, but their levels were found to be rather low or even negligible (Demeyer et al. 2000); findings in dry-cured ham were similar (Armenteros et al. 2012). In other cases, the generation of N-nitrosamines seems to be due to the reaction of nitrite remaining in the meat product with amine additives present in rubber nettings (Sen et al. 1987).

Potassium and sodium salts of nitrite (E 249 and E 250) and nitrate (E-251 and E-252) are authorized for use up to certain levels in several foodstuffs such as non-heat-treated, cured and dried meat products, other cured meat products, canned meat products, and bacon. This authorization is based largely on the proven inhibitory effect of nitrite on *Clostridium botulinum*. Thus, nitrate and nitrite can be used as effective preservatives, but the amounts used must be limited to those strictly necessary for microbiological safety assurance to reduce the potential generation of nitrosamines (European Food Safety Authority 2003). Nitrites and nitrates were authorized as additives in Directive 95/2/EC on food additives other than colors and sweeteners. This directive was amended by Directive 2006/52/EC of 5 July 2006, where the initial amounts were replaced by maximum levels to be added. In general, the maximum amount of nitrite that can be added to all meat products is 150 mg/kg, whereas nitrate can be added in the case of unheated meat products to a maximum of 150 mg/kg (Honikel 2010). There are some exceptions like Wiltshire or dry-cured bacon, where the amounts are slightly higher.

Nitrate and nitrite play other roles in meat and poultry; they confer an antioxidant benefit, protect lipids from oxidation, and improve product aroma and color (Toldrá et al. 2009).

Several types of extraction like steam distillation, liquid-liquid extraction, solvent extraction, SPE, or supercritical fluid extraction can be used for the separation of nitrosamines from meat matrices (Fiddler and Pensabene 1996; Raoul et al. 1997; Rath and Reyes 2009). Once extracted, volatile N-nitrosamines, or nonvolatile nitrosamines previously derivatized by acylation or trimethylsilylation, are usually analyzed by GC coupled to a thermal energy analyzer or mass spectrometry detectors. LC-MS and MS-MS in the mode of atmospheric pressure chemical ionization is used for the analysis of nonvolatile nitrosamines (Eerola et al. 1998; Rath and Reyes 2009).

1.6.2 Heterocyclic Amines

Heterocyclic amines (HAs) are formed by reaction of amino acids, alone or with creatine or creatinine, when meat is cooked at high temperatures. Thus, high levels of HAs may be found in well-done fried, broiled, and grilled/barbecued meats and meat products (Sinha et al. 1998), whereas lower levels of HAs are formed in oven roasting and baking at low temperatures. In general, HA generation is facilitated by the direct contact of meat with the heating source device, especially at surface temperatures over 150 °C, and the amounts increase exponentially with temperature (Felton et al. 2002). The content of creatine in raw meat and poultry is relatively high, within a range of 240–380 mg/100 g of meat depending on the type of muscle metabolism being higher in glycolytic muscles (Mora et al. 2008a). When meat is cooked or processed, creatine is progressively converted into creatinine (Mora et al. 2008b)

Two major classes of HAs are found in overcooked meat: aminoimidazol-quinolines and aminoimidazol-pyridines. The HAs most frequently found in meat (Table 1.19) are 2-amino-1-methyl-6-phenylimidazol(4,5,b)pyridine (PhIP) and 2-amino-3,8-dimethylimidazo(4,5,f)quinoxiline (MeIQx). Other minor compounds are 2-amino-9-H-pyrido(2,3,b)indole (AC); 2-amino-3,4-dimethylimidazo(4,5,f)quinoline (IQ); and 2-amino-3,4,8-trimethylimidazo(4,5,f)quinoxiline (DiMeIQx) (Jaksyn et al. 2004). The intake of these HAs has been related to certain types of cancer (Bogen 1994; Augustsson et al. 1999). In fact, the intake of HAs may follow a genotoxic mechanism, leading to DNA binding, mutation, and cancer initiation (Felton et al. 2002). Mutagenic analysis of cooked meat has shown that approximately 35 % of the total mutagenicity was due to MeIQx, usually present at 1 μg/kg original fresh weight of beef. Other mutagens were 4,8 DiMeIQx, present at 0.5 μg/kg, and PhIP, present at 15 μg/kg. This last amine, PhIP, has been reported in beef at levels tenfold higher than other HAs (Felton et al. 1986). Other minor mutagens were IQ (0.02 μg/kg), MeIQ (<0.01 μg/kg), and TMIP (0.5 μg/kg) (Felton et al. 1984). In any case, the assessment of HA intake is rather difficult because its content in meat depends on the type of cooking, temperature, and time (Bjeldanes et al. 1983).

The analysis of HAs is rather complex. Extraction is performed by aqueous extraction at pH 2, followed by absorption and elution with a XAD-2 resin (Bjeldanes et al. 1982) or using SPE. The samples can be analyzed by LC or GC coupled to mass spectrometry (Felton et al. 2002). NMR may also be used.

1.6.3 Polycyclic Aromatic Hydrocarbons

Smoking has a very long history of use in meat preservation. Smoke is generated by the controlled combustion of certain natural hard woods, sometimes accompanied by aromatic herbs and spices or even moist wood chips. It also gives to the meat product a characteristic smoky flavor, attributable to some flavoring substances. The smoke is condensed and adsorbed on the surface of the meat product, but its

Table 1.19 List of heterocyclic amines with carcinogenic properties

Heterocyclic amines	Structure	CAS number	Molecular mass (g/mol)
2-amino-1-methyl-6-phenylimidazo (4,5,b)pyridine (PhIP)	N, N, NH 2, N, Me, Ph	105650-23-5	224.24
2-amino-3,4-dimethylimidazo (4,5,f)quinoxiline (MeIQ)	NH 2, N, N, Me, N, Me	77094-11-2	212.25

2-amino-3,8-dimethylimidazo (4,5,f)quinoxiline (8-MeIQx)		77500-04-0	213.11
2-amino-9-H-pyrido (2,3,b)indole (AC)		26148-68-5	183.21
2-amino-3,4-dimethylimidazo (4,5,f)quinoline (IQ)		95896-78-9	227.27

penetration rate depends on several factors closely related to the process technology like temperature, humidity, volatility, and velocity of the smoke. Further information on smoking, its production, and application is widely described elsewhere (Sikorski and Kolakowski 2010).

Despite the pluses of smoking meat products, smoke also contains some health-hazardous compounds like polycyclic aromatic hydrocarbons (PAHs), phenols, and formaldehyde (Bem 1995). PAHs are generated by incomplete burning of wood especially within a temperature range of 500–700 °C and when the oxygen supply is limited (Simko 2009a). The Scientific Committee on Food of the European Union assessed 33 PAHs in 2002 and identified 15 with genotoxic and carcinogenic properties (Table 1.20) as having a high priorty. The determination of all PAHs is quite complex and the committee proposed benzo-a-pyrene (BaP), which also possesses carcinogenic properties, as a marker. The maximum levels for PAHs in certain foods was set by Regulation 466/2001 as amended by Regulation 208/2005 (European Commission 2005). BaP is used as an indicator of the presence of PAHs in meat, and the EC regulation limited its amount to 5 μg/kg in smoked meat and smoked meat products.

Most PAHs have been classified as 2A by the International Agency of Research on Cancer. Formaldehyde can promote cancerous tumors, whereas some smoke phenols can react to form highly toxic reaction products like nitrosophenols, nitrophenols, polymeric nitroso compounds, and other toxic compounds or even catalyze the formation of nitrosamines (Bem 1995). Meat products that are extensively smoked in old or inadequate smokehouses are the most dangerous because the PAH levels there can reach amounts near 100 μg/kg (Simko 2009a). When technology is correctly applied, the PAH content is below 1 μg/kg. Information about PAH content in 313 food items in 23 countries was published a few years ago (Jaksyn et al. 2004). In any case, the content in PAHs is highly variable because it depends on the type of technology and its processing variables like the use of direct or indirect smoking, the type of generator used, the type and composition of wood and herbs, accessibility to oxygen, and the temperature and time of the process.

The presence of substantial amounts of PAHs in smoked meat products prompted the development of alternative processes to reduce contamination with hazardous substances. Such reduction of PAHs in smoked meat products could be achieved through the filtration of particles, use of cooling traps, application of lower temperatures, or reduction of the process duration. An alternative strategy, most commonly applied today, consists in the application of liquid smoke on the surface of a meat product. Such liquid smoke flavorings can be added to various foods, within a range of 0.1–1.0 %, to replace the smoking process or to impart a smoke flavor to foods that are not traditionally smoked. Smoke flavorings are produced by controlled thermal degradation of wood in the presence of a limited supply of oxygen (pyrolysis), subsequent condensation of the vapors, and fractionation of the resulting liquid products. Then the primary products, which are the primary smoke condensates and the primary tar fractions, may be further processed to produce smoke flavorings applied on the foods (European Food Safety Authority2005). But primary products may contain a wide variety of compounds including PAHs (Jennings 1990; Maga 1987),

Table 1.20 List of polycyclic aromatic compounds (PAHs), with known carcinogenic or genotoxic properties as identified by the Scientific Committee of Food that may be potentially present in primary products used for production of smoke flavorings (EFSA 2005)

Polycyclic aromatic compounds (PAHs)	Chemical structure	CAS number	Molecular mass (g/mol)
Benz[a]anthracene		56-55-3	228.29
Benzo[b]fluoranthene		205-99-2	252.31
Benzo[j]fluoranthene		205-82-3	252.31
Benzo[k]fluoranthene		207-08-9	252.31
Benzo[g,h,i]perylene		191-24-2	276.33
Benzo[a]pyrene		50-32-8	252.31
Chrysene		218 01-9	228.29
Cyclopenta[c,d]pyrene		27208-37-3	226.27
Dibenz[a,h]anthracene		53-70-3	278.35
Dibenzo[a,e]pyrene		192-65-4	302.37
Dibenzo[a,h]pyrene		192-51-8	302.37
Dibenzo[a,i]pyrene		189-55-9	302.37

(continued)

Table 1.20 (continued)

Polycyclic aromatic compounds (PAHs)	Chemical structure	CAS number	Molecular mass (g/mol)
Dibenzo[a,l]pyrene	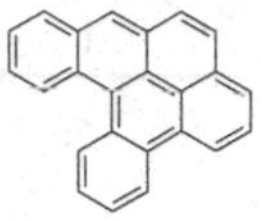	191-30-0	302.37
Indeno[1,2,3-cd]pyrene	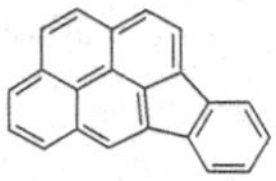	193-39-5	276.33
5-Methylchrysene	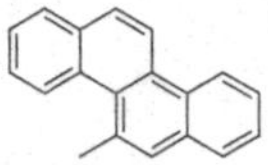	3697-24-3	242.31

even though their toxicological effects can vary significantly among preparations because of the type of production process, the qualitative and quantitative composition, the concentration used in the flavoring, and the final use levels (Scientific Committee for Food 1995). Smoke flavoring of primary products is evaluated by the European Food Safety Authority (EFSA) in accordance with a guidance document where main relevant data (technical data, proposed uses, dietary exposure assessment, and toxicological data) must be provided (European Food Safety Authority 2005). The use of smoke flavoring in primary products is controlled in the European Union through Council Regulation 2065/2003 (European Commission 2003) on smoke flavorings used or intended for use in or on foods. Under this regulation, the use of a primary product in and on foods shall only be authorized if it is sufficiently demonstrated that it does not present risks to human health. It lays down a procedure for the evaluation and authorization of primary smoke condensates and primary tar fractions and for the establishment of a list of primary smoke condensates and tar fractions to the exclusion of all others and their conditions of use.

According to this regulation (European Commission 2003), the maximum amounts of BaP and benzo-a-anthracene allowed in liquid smoke flavoring in primary products is 10 and 20 μg/kg, respectively. The list of primary products that are allowed for use as such in or on food or for the production of derived smoke flavorings is issued by the EFSA based on the available studies on subchronic toxicity and genotoxicity.

Regulation 627/2006 (European Commission 2006b) implemented Regulation 2065/2003 regarding quality criteria for validated analytical methods for sampling, identification, and characterization of primary smoke products. This regulation included methods of sampling, sample preparation, and criteria for methods of analysis; all these were essential for having available techniques by which one could reliably analyze the 15 priority PAHs. To obtain reliable data for official food controls, the European Commission assigned a Community Reference Laboratory to PAHs in 2006 (Wenzl et al. 2006). The detection of PAH compounds can be performed with either GC coupled to a flame ionization detector or HPLC coupled to ultraviolet or fluorescence detectors. Identification and confirmation of PAHs may be performed using mass spectrometry detectors coupled to either GC or HPLC. A detailed description of methods of analysis for the detection and identification of PAHs in meat products was recently published (Simko 2009b).

1.6.4 Biogenic Amines in Fermented Meats and Poultry

The generation of biogenic amines is brought about through the action of microbial decarboxylase activity against precursor amino acids. This generation is usually observed in fermented foods, either because of microbial contamination or the use of a microbial starter having such decarboxylase activity. Some lactic acid bacteria – enterococci and staphylococci – are able to generate tyramine and phenylethylamine (Bover-Cid et al. 2001; Straub et al. 1995). Tyramine is the most commonly found amine in fermented sausages and cadaverine and putrescine, though with more variability and at lower levels; histamine is rarely present, and the contents of phenylethylamine and tryptamine are usually low (Vidal-Carou et al. 2007). Table 1.21 summarizes the different amines and their respective amino acid precursors. Based on their chemical structure, amines can be classified as aromatic amines (histamine, tyramine, phenylethylamine, and tryptamine), aliphatic diamines (putresine and cadaverine), and aliphatic polyamines (agmatine, spermidine, and spermine). In general, the consumption of low amounts of amines in fermented meats does not pose a risk for humans because the ingested amines are oxidatively deaminated by the enzyme monoamine oxidase (MAO). Trouble can appear when large amounts of amines are consumed or for those consumers taking medicines containing MAO inhibitors. Symptoms such as migraine or hypertensive crisis may appear due to their vasoactive and psychoactive properties (Shalaby 1996). For instance, the estimated tolerance level for tyramine is 100–800 mg/kg (Nout 1994); among other symptoms, tyramine can cause the release of stored monoamines such as dopamine, norepinephrine, and epinephrine.

The presence of amines constitutes a good indicator of the hygienic quality of meat, especially when either cadaverine or putrescine are present, that would indicate the presence of contaminating meat flora (Bover-Cid et al. 2000). In fact, a biogenic amine index to measure the freshness of meat and its hygienic quality, as is already used for fish, has been proposed. Several proposals for this index could be based on particular amines like cadaverine for meat and poultry (Vinci and Antonelli 2002), tyramine and putrescine for chicken (Patsias et al. 2006), or tyramine, cadaverine, putrescine, and histamine for cooked pork (Hernández-Jover et al. 1996). However, the most common problems arise in connection with fermented products. In these cases, the presence of amines is due to the decarboxylase activity in any of the microorganisms present as natural flora or in added culture starters (Eerola et al. 1996).

Preventive measures to avoid the generation of biogenic amines are relatively easy to follow. The selection of raw materials with correct hygienic conditions and

Table 1.21 Amines, their main characteristics, and amino acid of origin

Amines	Structure	CAS number	Molecular mass (g/mol)	Amino acid of origin
Tyramine	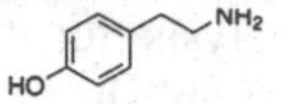	51-67-2	137.18	Tyrosine
Phenylethylamine	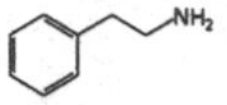	64-04-0	121.18	Phenylalanine
Histamine	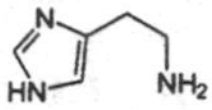	51-45-6	111.15	Histidine
Tryptamine	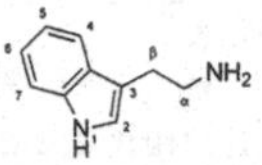	61-54-1	160.22	Tryptophane
Cadaverine		462-94-2	102.18	Lysine
Putrescine	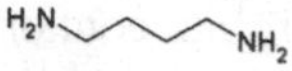	110-60-1	88.15	Ornithine
Agmatine	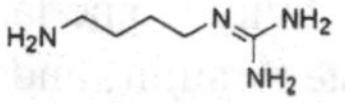	306-60-5	130.19	Arginine
Spermidine	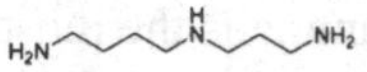	124-20-9	145.25	Putrescine
Spermine	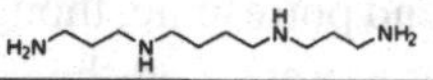	71-44-3	202.34	Putrescine

good manufacturing practices are of primary importance, as is, of course, the screening of starter cultures for any decarboxylase activity and even the use of starter cultures having amine oxidase activity (Talon et al. 2002; Vidal-Carou et al. 2007).

Analysis of biogenic amines includes a liquid extraction with acid solutions or organic solvents followed by cleanup of the extract. Solvents containing trichloroacetic acid or perchloric acid are widely used because they also contribute to protein precipitation. Once centrifuged and filtered, amines are then analyzed by HPLC with either ion exchange or reversed phase with ion pairs followed by ultraviolet–visible or fluorescence detection. The response of amines to detection systems is rather poor and requires either pre- or postcolumn derivatization to increase their sensitivity. Many derivatization agents exist, but dansyl chloride and o-phthalaldehyde (OPA) are the most commonly used ones. Sample pretreatment for OPA is easier and has some additional advantages such as the possibility for full automation and better sensitivity through fluorescence detection. Additional details for analysis are given elsewhere (Vidal-Carou et al. 2009; Ruiz-Capillas and Jiménez-Colmenero 2010).

An enzyme sensor employing diamine oxidase immobilized on a preactivated immunodyne membrane in combination with an oxygen electrode was recently developed and optimized to estimate the content of total amines in dry-fermented sausages. The measurements of the enzyme sensor were well correlated to those obtained using a standard HPLC method and could constitute a reliable screening method to detect the presence of biogenic amines in dry-fermented sausages (Hernández-Cázares et al. 2011). Other methods to measure meat freshness are based on the detection of nucleoside generation, basically hypoxanthine. Thus, pork meat freshness was successfully evaluated with an enzyme sensor using immobilized xanthine oxidase to detect hypoxanthine and xanthine (Hernández-Cázares et al. 2010). Other methods that have been developed for the evaluation of fish freshness have used a potentiometric sensor (Barat et al. 2008; Gil et al. 2008)

1.6.5 Lipid Oxidation Products

Lipid oxidation involves the degradation of polyunsaturated fatty acids (PUFAs), vitamins, and other tissue components and the generation of free radicals, which lead to the development of rancid odors and changes in color and texture in foodstuffs (Kanner 1994). Lipid oxidation is a cause of major deterioration in meat and meat products. It has been extensively studied, and its impact on meat quality through the formation of rancid odors, deterioration of flavor, and associated serious health concerns is well known (Kanner 1994; Byrne et al. 2001, 2002; Elmore et al. 2000).

Lipid oxidation concerns mainly triacylglycerols, phospholipids, lipoproteins, and cholesterol. Phospholipids are very susceptible to oxidation due to their high content of polyunsaturated fatty acids. Oxidation may be catalyzed by light, metal ions (e.g., iron, copper, cobalt, manganese), or enzymes. When oxidation is catalyzed by lipoxygenase, preformed hydroperoxide activates the enzyme (Honikel 2009). Another catalyzer of lipid oxidation in fermented meats is hydrogen peroxide, which is generated by peroxide-forming bacteria during meat fermentation.

Lipid oxidation follows a free radical mechanism consisting of three steps: initiation, propagation, and termination. The primary products of oxidation are hydroperoxides, which are relatively unstable and odorless. The secondary products of oxidation, such as aldehydes, ketones, alkanes, alkenes, alcohols, esters, acids, and hydrocarbons, can contribute to off-flavors, color deterioration, and potential generation of toxic compounds (Kanner 1994). Some of these may be chronic toxicants, especially when formed in large amounts because they can contribute to aging, cancer, and cardiovascular diseases (Hotchkiss and Parker 1990). The rancid taste typically associated with lipid oxidation is mainly to aldehydes that have low threshold values.

Several methods exist for measuring lipid oxidation in meat products. TBARS consists in the spectrophotometric determination of malondialdehyde (MDA) formation as an index of oxidative status. It is the most commonly used method, even though it is not specific and is somewhat error prone. An interesting alternative is the analysis of aldehydes, especially hexanal, by static headspace GC, dynamic headspace GC, or solid-phase microextraction GC (Ross and Smith 2006).

Cholesterol oxidation may occur through an autoxidative process or in conjunction with fatty acid oxidation, especially when reheating chilled meat or during the chilled storage of meat (Hotchkiss and Parker 1990). Cholesterol oxides are considered to be harmful to human health due to its role in the buildup of arteriosclerotic plaque, but they can also be mutagenic, carcinogenic, and cytotoxic (Guardiola et al. 1996). No cholesterol oxides were detected after heating pork sausages (Baggio and Bragagnolo 2006). However, other studies conducted on European sausages detected up to 1.5 μg/g of cholesterol oxides despite a low 0.17 % of cholesterol oxidation (Demeyer et al. 2000). These values were below the toxic levels observed through in vivo tests with laboratory animals (Bösinger et al. 1993). The major cholesterol oxide found in an Italian sausage was 7-ketocholesterol, whereas α-5, 6-epoxycholesterol was the major end product in other analyzed sausages (Demeyer et al. 2000).

1.6.6 *Protein Oxidation Products*

Oxidation of proteins constitutes a major threat to meat quality because it can lead to organoleptic quality degradation of meat products and thus affect flavor and color and cause serious health concerns (Xiong 2000; Byrne et al. 2001, 2002). The oxidation of meat proteins also has an impact on the nutritional value of meat because it involves the loss of essential amino acids and decreases protein digestibility (Xiong 2000). Despite these facts, little attention has been paid to protein oxidation in meat and meat products (Elias et al. 2008).

Muscle proteins may be oxidized by reactive oxygen species, for instance, by certain bacteria that generate hydrogen peroxide during meat fermentation. In other cases, metal ions or lipid oxidation may promote the oxidative damage of proteins through the prooxidant activity of primary (hydroperoxides) and secondary (aldehydes, ketones) lipid oxidation products (Estévez et al. 2008). Protein oxidation mainly occurs via free radical reactions in which peroxyl radicals generated in the first stages of PUFA oxidation can abstract hydrogen atoms from protein molecules, leading to the formation of protein radicals. The formation of noncovalent complexes between lipid oxidation products and reactive amino acid residues, as well as the presence of some particular metal such as copper and iron, can also lead to protein radical generation (Viljanen et al. 2004). Protein oxidation may lead to a substantial reduction in eating quality such as reduced tenderness and juiciness, flavor deterioration, and discoloration in meat (Xiong 2000) and in dry-cured meat products (Armenteros et al. 2009).

Protein oxidation is responsible for many biological modifications such as protein fragmentation or aggregation, changes in hydrophobicity, and protein solubility, affecting technological properties such as gelation (Srinivasan and Xiong 1996), emulsification (Srinivasan and Hultin 1997), solubility, and water-holding capacity (Ooizumi and Xiong 2004). In addition, protein oxidation might also play a role in meat tenderness (Rowe et al. 2004a) by controlling protease activity (Rowe et al. 2004b) but also by reducing the susceptibility of myofibrillar proteins to proteolysis (Morzel et al. 2006).

The main modification of amino acids by oxidation, especially proline, arginine, lysine, methionine, and cysteine residues, consists of the formation of carbonyl derivatives (Giulivi et al. 2003; Gatellier et al. 2010). The formation of carbonyl compounds can be used as a kind of measurement of protein damage by oxygen radicals under processing conditions (Estévez 2011). In fact, there is a significant effect of cooking time and temperature on the formation of carbonyls. Ganhão et al. (2010) determined that cold storage had a significant effect on protein oxidation as the amount of carbonyl compounds increased significantly in porcine patties. Other oxidative mechanisms consist of thiol oxidation and aromatic hydroxylation (Morzel et al. 2006). Sulfur amino acids of proteins are more susceptible to oxidation by peroxide reagents like hydrogen peroxide. Thus, cystine is oxidized only partly to cysteic acid, whereas methionine is oxidized to methionine sulfoxide and methionine sulfone in small amounts (Slump and Schreuder 1973). Sulfinic and cysteic acids can also be produced by direct oxidation of cysteine (Finley et al. 1981). The oxidation of homocystine can generate homolanthionine sulfoxide as the main product (Lipton et al. 1977). Peptides such as reduced glutathione can also be oxidized by hydrogen peroxide. Oxidation rates increase with pH, and most of the cysteine in the glutathione is oxidized to the monoxide or dioxide forms.

A method used for the quantification of carbonyl compounds in meat and meat products is based on the derivatization of carbonyl protein groups with the 2,4-dinitrophenylhydrazine to form hydrazones, and then the absorbance is measured at 370 nm (Oliver et al. 1987). Another method to evaluate protein oxidation is based on the conjugated fluorophores resulting from reactions between lipid oxidation products (aldehydes) and amino groups. This fluorescence can be detected at excitation and emission wavelengths of 350 and 450 nm, respectively (Viljanen et al. 2004). But these methods are nonspecific and may give large margins of error. Recently, a method based on the measurement of α-aminoadipic and γ-glutamic semialdehydes (AAS and GGS, respectively) was considered as a good alternative to measure specific biomarkers of oxidative damage (Estévez et al. 2008). Both semialdehydes are formed as the main carbonyl products from metal-catalyzed oxidized proteins. This method uses LC-ESI mass spectrometry and was recently applied in a survey of protein oxidation in different meat products. The results showed that dry-cured ham and dry-cured sausages had the highest amount of GGS, followed by liver pâté and cooked sausages. Ground meat had the lowest GGS levels (Armenteros et al. 2009).

1.6.7 Irradiation-Derived Compounds

Meat and poultry may be exposed to ionizing radiation under controlled conditions for disinfection purposes. The main types of ionizing radiation that are used for food irradiation and that are internationally recognized for the treatment of foods are gamma rays, which is the most widely used, along with Co-60, e-beams, and X-rays. Food irradiation is regulated in the EU by Directive 1999/2/EC. The list of foods authorized for irradiation treatment in the whole EU is given in Directive 1999/3/EC. It also includes a list of 23 approved food-irradiation facilities in 12 member states (Belgium, Bulgaria, Czech Republic, Germany, Spain, France, Hungary, Italy, the Netherlands, Poland, Romania, and the UK). Member states must inform the European Commission every year about the amounts of food irradiated in their respective facilities, and the Commission publishes the corresponding annual data. Foodstuffs irradiated include dried aromatic herbs, spices and vegetable seasonings, fresh and dried vegetables, dried fruits, various dehydrated products, starch, poultry, meat, fish and shellfish, frog legs and frog parts, shrimp, egg white, egg powder, dehydrated blood, and Arabic gum (European Commission 2009). Furthermore, food irradiation is approved in more than 60 countries worldwide for use in a wide variety of foodstuffs.

Several chemical substances like hydrocarbons, furans, alkylcyclobutanones, cholesterol oxides, and aldehydes can be formed as a consequence of the ionizing radiation treatment of meat or poultry (Sommers et al. 2006), though they can also be generated when subjected to other processing treatments, except for 2-alkylcyclobutanones, which are considered unique radiolytic products. However, Variyar et al. (2008) detected 2-dodecylcyclobutanone (2-DCB) and 2-tetradecylcyclobutanone (2-tDCB) in commercial nonirradiated and fresh cashew nut samples, as well as 2-decylcyclobutanone and 2-DCB in nonirradiated nutmeg samples, but these results require confirmation.

The extent of the reactions induced by irradiation treatment are strongly dependent on treatment conditions such as absorbed dose, dose rate, presence or absence of oxygen, and temperature but also by the composition of meat and whether it is in a frozen or refrigerated state. The effects may be minimized by using low temperatures and reducing the presence of oxygen (Stefanova et al. 2010). The changes in nutrient composition induced by irradiation are relatively small. Some vitamins such as thiamine and vitamins E and A appear to be the most affected (Smith and Pillai 2004).

Ten validated methods were standardized by the European Committee for Standardisation (CEN) as European Standards (EN). They are (Stewart 2009) (1) biological, based on the ratio of living to dead microorganisms, DNA strand breakage, the direct epifluorescent filter technique/aerobic plate count or DNA comet assay; (2) physical, based on the technique of electron spin resonance spectroscopy, thermoluminescence, or photostimulated luminescence; and (3) chemical methods, based on the measurement of radiolytic products like radiolytic hydrocarbons and 2-alkylcyclobutanones that are extracted and then separated by GC and detected and identified using mass spectrometry. In the last case, the radiolytic products that are not present in nonirradiated foods are derived largely from the major fatty acids in meat and poultry (Table 1.22). The corresponding cyclobutanones that are formed are 2-dodecyl-cyclobutanone (2-dDCB), 2-tetradecylcyclobutanone (2-tDCB), 2-tetradec-5′-enyl-cyclobutanone (2-tDeCB), and 2-tetradeca-5′,8′-dienylcyclobutanone (2-tDdeCB) (Horvatovich et al. 2005). In fact, 2-dDCB and 2-tDCB constitute good markers for the detection of irradiated meat or poultry. Thus, the analysis of 2-dDCB was used to detect the presence of irradiated mechanically recovered meat in food preparations (Marchioni et al. 2002). Other authors have used solid-phase microextraction for the extraction of 2-DCB from irradiated ground beef (Caja et al. 2008) or a direct solvent extraction method for 2-DCB in irradiated chicken (Tewfik 2008a, b). 2-tDCB was also detected in irradiated chicken meat

Table 1.22 Main radiolytic compounds, characteristics, and fatty acid of origin

Hydrocarbons	Alkyl- cyclobutanones	Molecular mass (g/mol)	Fatty acid of origin
Didecene Tridecane	2-decyl-cyclobutanone (2-DCB)	210.36	Myristic acid
Tetradecene Pentadecane	2-Dodecyl-cyclobutanone (2-dDCB)	238.41	Palmitic acid
Hexadecene Heptadecane	2-Tetradecyl-cyclobutanone (2-tDCB)	266.46	Stearic acid
Tetradecadiene Hexadecene	2-(dodec-5′-enyl)-cyclobutanone (2-dDeCB)	236.39	Palmitoleic acid
Heptadecene Hexadecadiene	2-Tetradeca-5′-enyl-cyclobutanone (2-tDeCB)	264.45	Oleic acid
Heptadecadiene Hexadecatriene	2-Tetradeca-5′-8′-dienyl-cyclobutanone (2-tD2eCB)	262.44	Linoleic acid
Heptadecatriene Hexadecatetraene	2-(tetradeca-5′,8′,11′-trienyl)-cyclobutanone (2-tD3eCB)	260.42	Linolenic acid

(Stewart et al. 2001; Zanardi et al. 2007). The levels of detection are as low as 0.03–0.05 μg/g 2-DCB per kilogray in irradiated ground beef (Gadgil et al. 2002, 2005) or 0.1 μg/g per kilogray in irradiated lyophilized poultry meat after 28 days under refrigerated storage (Horvatovich et al. 2005).

References

Ahlborg UG, Becking GC, Birnbaum LS, Brouwer A, Derks HJGM, Feely M, Golor G, Hanberg A, Larsen JC, Liem AKD, Safe SH, Schlatter C, Waern F, Younes M, Yrjänheikki E (1994) Toxic equivalency factors for dioxin-like PCBs. Chemosphere 28:1049–1106

Antignac JP, de Wasch K, Monteau F, De Brabander H, Andre F, Le Bizec B (2005) The ion suppression phenomenon in liquid chromatography-mass spectrometry and its consequences in the field of residue analysis. Anal Chim Acta 529:129–136

Armenteros M, Heinonen M, Ollilainen V, Toldrá F, Estévez M (2009) Analysis of protein carbonyls in meat products by using the DNPH method, fluorescence spectroscopy and liquid chromatography-electrospray ionisation-mass spectrometry (LC-ESI-MS). Meat Sci 83:104–112

Armenteros M, Aristoy MC, Toldrá F (2012) Evolution of nitrate and nitrite during the processing of dry-cured ham with partial replacement of NaCl by other chloride salts. Meat Sci. DOI 10.1016/j.meatsci.2012.02.017

Ashwin HM, Stead SL, Taylor JC, Startin JR, Richmond SF, Homer V, Bigwood T, Sharman M (2005) Development and validation of screening and confirmatory methods for the detection of chloramphenicol and chloramphenicol glucuronide using SPR biosensor and liquid chromatography-tandem mass spectrometry. Anal Chim Acta 529:103–108

Augustsson K, Skog K, Jagerstad M, Dickman PW, Steineck G (1999) Dietary heterocyclic amines and cancer of the colon, rectum, bladder, and kidney: a population-based study. Lancet 353:686–687

Baggiani C, Anfossi L, Giovannoli C (2007) Solid phase extraction of food contaminants using molecular imprinted polymers. Anal Chim Acta 591:29–39

Baggio SR, Bragagnolo N (2006) The effect of heat treatment on the cholesterol oxides, cholesterol, total lipid and fatty acid contents of processed meat products. Food Chem 95:611–619

Barat JM, Gil L, García-Breijo E, Aristoy MC, Toldrá F, Martínez-Máñez R, Soto J (2008) Freshness monitoring of sea bream (*Sparus aurata*) with a potentiometric sensor. Food Chem 108:681–688

Barbosa J, Cruz C, Connolly L, Elliott CT, Lovgren T, Tuomola M (2005) Food poisoning by clenbuterol in Portugal. Food Addit Contam 22:563–566

Bastide NM, Pierre FHF, Corpet DE (2011) Heme iron from meat and risk of colorectal cancer: a meta-analysis and a review of the mechanisms involved. Cancer Prev Res 4:177–184

Bem Z (1995) Desirable and undesirable effects of smoking meat products. Die Fleischerei 3:3–8

Berggren C, Bayoudh S, Sherrington D, Ensing K (2000) Use of molecularly imprinted solid-phase extraction for the selective clean-up of clenbuterol from calf urine. J Chromatogr A 889: 105–110

Bergwerff AA (2005) Rapid assays for detection of residues of veterinary drugs. In: van Amerongen A, Barug D, Lauwars M (eds) Rapid methods for biological and chemical contaminants in food and feed. Wageningen Academic Publishers, Wageningen, the Netherlands, pp 259–292

Bergwerff AA, Schloesser J (2003) Residue determination. In: Caballero B, Trugo L, Finglas P (eds) Encyclopedia of food sciences and nutrition, 2nd edn. Elsevier, London, pp 254–261

Bienemann-Ploum M, Korpimaki T, Haasnoot W, Kohen F (2005) Comparison of multi-sulfonamide biosensor immunoassays. Anal Chim Acta 529:115–122

Bjeldanes LF, Grose KR, Davis PH, Stuermer DH, Healy SK, Felton JS (1982) An XAD-2 resin method for efficient extraction of mutagens from fried ground beef. Mutat Res 105:43–49

Bjeldanes LF, Morris MM, Timourian H, Hatch FT (1983) Effects of meat composition and cooking conditions on mutagen formation in fried ground beef. J Agric Food Chem 31:18–21

Bogen KT (1994) Cancer potencies of heterocyclic amines found in cooked foods. Food Chem Toxicol 32:505–515

Borràs S, Companyó R, Granados M, Guiteras J, Pérez-Vendrell AM, Brufau J, Medina M, Bosch J (2011a) Analysis of antimicrobial agents in animal feed. Trends Anal Chem 30:1042–1064

Borràs S, Companyó R, Guiteras J (2011b) Analysis of sulfonamides in animal feeds by liquid chromatography with fluorescence detection. J Agric Food Chem 59:5240–5247

Bösinger S, Luf W, Brandl E (1993) Oxysterols: their occurrence and biological effects. Int Dairy J 3:1–33

Bover-Cid S, Izquierdo-Pulido M, Vidal-Carou MC (2000) Influence of hygienic quality of raw materials on biogenic amine production during ripening and storage of dry fermented sausages. J Food Prot 63:1544–1550

Bover-Cid S, Hugas M, Izquierdo-Pulido M, Vidal-Carou MC (2001) Amino acid-decarboxylase activity of bacteria isolated from fermented pork sausages. Int J Food Microbiol 66:185–189

Boyd B, Bjork H, Billing J, Shimelis O, Axelsson S, Leonora M, Yilmaz E (2007) Development of an improved method for trace analysis of chloramphenicol using molecularly imprinted polymers. J Chromatogr A 1174:63–71

Brockman RP, Laarveld R (1986) Hormonal regulation of metabolism in ruminants. A review. Livest Prod Sci 14:313–317

Butaye P, Devriese LA, Haesebrouck F (2001) Differences in antibiotic resistance patterns of *Enterococcus faecalis* and *Enterococcus faecium* strains isolated from farm and pet animals. Antimicrob Agents Chemother 45:1374–1378

Byrne DV, Bredie WLP, Bak LS, Bertelsen G, Martens H, Martens M (2001) Sensory and chemical analysis of cooked porcine meat patties in relation to warmed-over flavour and pre-slaughter stress. Meat Sci 59:229–249

Byrne DV, Bredie WLP, Mottram DS, Martens M (2002) Sensory and chemical investigations on the effect of oven cooking on warmed-over flavor development in chicken meat. Meat Sci 61:127–139

Byrnes SD (2005) Demystifying 21 CFR Part 556—tolerances for residues of new animal drugs in food. Regul Toxicol Pharmacol 42:324–327

Caja MM, del Castillo MLR, Blanch GP (2008) Solid phase microextraction as a methodology in the detection of irradiation markers in ground beef. Food Chem 110:531–537

Campbell HM, Armstrong JF (2007) Determination of zearalenone in cereal grains, animal feed, and feed ingredients using immunoaffinity column chromatography and liquid chromatography: interlaboratory study. J AOAC Int 90:1610–1622

Cassens RG (1997) Composition and safety of cured meats in the USA. Food Chem 59:561–566

Cerniglia CE, Kotarski S (1998) Evaluation of veterinary drug residues in food for their potential to affect human intestinal microflora. Regul Toxicol Pharmacol 29:238–261

Cerniglia CE, Kotarski S (2005) Approaches in the safety evaluations of veterinary antimicrobial agents in food to determine the effects on the human intestinal microflora. J Vet Pharmacol Ther 28:3–20

National Archives and Records Administration (2008) Tolerances for residues of new animal drugs in food. Code of Federal Regulations, Title 21 Food and Drugs, Chapter I, Subchapter E, Part 556. http://ecfr.gpoaccess.gov/cgi/t/text. (Accessed 3 Jun 2008)

National Archives and Records Administration (2010) Tolerances and exemptions for pesticide chemical residues in food. Code of Federal Regulations, Title 40 Protection of Environment, Chapter I, Subchapter E, Part 180, Pesticide programs.http://ecfr.gpoaccess.gov/cgi/t/text/text-idx?c=ecfr&tpl=%2Findex.tpl. (Accessed 30 Mar 2012)

Chadwick RW, George SE, Claxton LD (1992) Role of gastrointestinal mucosa and microflora in the bioactivation of dietary and environmental mutagens or carcinogens. Drug Metab Rev 24:425–492

Cháfer-Pericás C, Maquieira Á, Puchades R (2010) Fast screening methods to detect antibiotic residues in food samples. TrAC Trends Anal Chem 29:1038–1049

Chiaochan C, Koesukwiwat U, Yudthavorasit S, Leepipatpiboon N (2010) Efficient hydrophilic interaction liquid chromatography–tandem mass spectrometry for the multiclass analysis of veterinary drugs in chicken muscle. Anal Chim Acta 682:117–129

Chifang P, Chuanlai X, Zhengyu J, Xiaogang C, Liying W (2006) Determination of anabolic steroid residues (medroxyprogesterone acetate) in pork by ELISA and comparison with liquid chromatography tandem mass spectrometry. J Food Sci 71:C044–C050

Cinquina AL, Longo F, Anastasi G, Giannetti L, Cozzani R (2003) Validation of a high perfor mance liquid chromatography method for the determination of oxytetracycline, tetracycline, chlortetracycline and doxycycline in bovine milk and muscle. J Chromatogr A 987:227–233

Connolly L, Thompson CS, Haughey SA, Traynor IM, Tittlemeier S, Elliot C (2007) The development of a multi.nitorimidazole residue analysis assay by optical biosensor via a proof of concept project to develop and assess a prototype test kit. Analytica Chimica Acta 598:155–161

Cooper KM, Ribeiro L, Alves P, Vozikis V, Tsitsamis S, Alfredssonk G, Lovgren T, Tuomola M, Takaloyy H, Iitiayy A, Sterkzz SS (2003) Interlaboratory ring test of time-resolved fluoroimmunoassays for zeranol and a-zearalenol and comparison with zeranol test kits. Food Additives and Contaminants 20:804–812

Cooper KM, Caddell A, Elliott CT, Kennedy DG (2004) Production and characterisation of polyclonal antibodies to a derivative of 3-amino-2-oxazolidinone, a metabolite of the nitrofuran furazolidone. Anal Chim Acta 520:79–86

Cooper KM, Samsonova JV, Plumpton L, Elliott CT, Kennedy DG (2007a) Enzyme immunoassay for semicarbazide—the nitrofuran metabolite and food contaminant. Anal Chim Acta 592:64–71

Cooper J, Delahaut P, Fodey TL, Elliott CT (2007b) Development of a rapid screening test for veterinary sedatives and the beta-blocker carazolol in porcine kidney by ELISA. Analyst 129:169–174

Cronly M, Behan P, Foley B, Malone E, Earley S, Gallagher M, Shearan P, Regan L (2010) Development and validation of a rapid multi-class method for the confirmation of fourteen prohibited medicinal additives in pig and poultry compound feed by liquid chromatography-tandem mass spectrometry. J Pharm Biomed Anal 53:929–938

Cross AJ, Ferrucci LM, Risch A, Graubard BI, Ward MH, Park Y, Hollenbeck AR, Schatzkin A, Sinha R (2010) A large prospective study of meat consumption and colorectal cancer risk: An investigation of potential mechanisms underlying this association. Cancer Res 70:2406–2414

Croubels S, Daeselaire E, De Baere S, De Backer P, Courtheyn D (2004) Feed and drug residues. In: Jensen W, Devine C, Dikemann M (eds) Encyclopedia of meat sciences. Elsevier, London, pp 1172–1187

De Brabander HF, Noppe H, Verheyden K, Vanden Bussche J, Wille K, Okerman L, Vanhaecke L, Reybroeck W, Ooghe S, Croubels S (2009) Residue analysis: future trends from a historical perspective. J Chromatogr A 1216:7964–7976

De Wasch K, Okerman L, Croubels S, De Brabander H, Van Hoof J, De Backer P (2001) Detection of residues of tetracycline antibiotics in pork and chicken meat: correlation between results of screening and confirmatory tests. Analyst 123:2737–2741

Demeyer DI, Raemakers M, Rizzo A, Holck A, De Smedt A, Ten Brink B, Hagen B, Montel C, Zanardi E, Murbrek E, Leroy F, Vanderdriessche F, Lorentsen K, Venema K, Sunesen L, Stahnke L, De Vuyst L, Talon R, Chizzolini R, Eerola S (2000) Control of bioflavor and safety in fermented sausages: first results of a European project. Food Res Int 33:171–180

Dixon SN (2001) Veterinary drug residues. In: Watson DH (ed) Food chemical safety. Vol 1: Contaminants. Woodhead, Cambridge, UK, pp 109–147

Draisci R, delli Quadri F, Achene L, Volpe G, Palleschi L, Palleschi G (2001) A new electrochemical enzyme-linked immunosorbent assay for the screening of macrolide antibiotic residues in bovine meat. Analyst 126:1942–1946

Dumont V, Huet AC, Traynor I, Elliott C, Delahaut P (2006) A surface plasmon resonance biosensor assay for the simultaneous determination of thiamphenicol, florefenicol, florefenicol amine and chloramphenicol residues in shrimps. Anal Chim Acta 567:179–183

Eerola S, Maijala R, Roig-Sangués AX, Salminen M, Hirvi T (1996) Biogenic amines in dry sausages as affected by starter culture and contaminant amine-positive Lactobacillus. J Food Sci 61:1243–1246

Eerola S, Otegui I, Saari L, Rizzo A (1998) Application of liquid chromatography atmospheric pressure chemical ionization mass spectrometry and tandem mass spectrometry to the determination of volatile nitrosamines in dry sausages. Food Addit Contam 15:270–279

European Community (1988) Council Directive 88/146/EEC of 7 March 1988 prohibiting the use in livestock farming of certain substances having a hormonal action. Off J Eur Union L 070:16

European Community (1993a) Commission Decision 93/256/EEC of 14 May 1993 laying down the methods to be used for detecting residues of substances having hormonal or a thyreostatic action. Off J Eur Union L 118:64

European Community (1993b) Commission Decision 93/256/EEC of 15 April 1993 laying down the reference methods and the list of the national reference laboratories for detecting residues. Off J Eur Union L 118:73

European Community (1996) Council Directive 96/23/EEC of 29 April 1996 on measures to monitor certain substances and residues thereof in live animals and animal products. Off J Eur Union L 125:10

European Community (2001) Council Regulation 2375/2001 of 29 November 2001 amending Commission Regulation (EC) No. 466/2001 setting maximum levels for certain contaminants in foodstuffs. Off J Eur Union L 321: 1

European Community (2002a) Commission Directive 2002/32/EC of 7 May 2002 on undesirable substances in animal feed. Off J Eur Union L 140:10

European Community (2002b) Commission Decision 2002/657/EEC of 17 August 2002 implementing Council Directive 96/23/EC concerning the performance of analytical methods and the interpretation of results. Off J Eur Union L 221:8

European Community (2003) Regulation No. 2065/2003 of the European Parliament and of the Council of 10 November 2003 on smoke flavourings used or intended for use in or on foods. Off J Eur Union L 309:1

European Community (2005) Commission Directive 2005/87/EC of 5 December 2005 amending Annex I to Directive 2002/32/EC of the European Parliament and of the Council on undesirable substances in animal feed as regards lead, fluorine and cadmium. Off J Eur Union L 318:19

European Community (2006a) Commission Regulation 1881/2006 of 19 December 2006 setting maximum levels for certain contaminants in foodstuffs. Off J Eur Union L 364:5–24

European Community (2006b) Commission Regulation 627/2006 of 21 April 2006 implementing Regulation (EC) No. 2065/2003 of the European Parliament and of the Council as regards quality criteria for validated analytical methods for sampling, identification and characterisation of primary smoke products. Off J Eur Union L 109:3–6

European Community (2009) Report from the Commission on Food Irradiation for the year 2007. Off J Eur Union C 242/02:2–18

European Food Safety Authority (2005) Guidance from the Scientific Panel on Food Additives, Flavourings, Processing Aids and Materials in Contact with Food. Guidance on submission of a dossier on a smoke flavouring primary product for evaluation by EFSA, adopted 7 October 2004; revised 27 April 2005. http://www.efsa.europa.eu/en/esfaJ/pub/492.htm. Accessed 10 Feb 2012

European Food Safety Authority (2003) The effects of nitrites/nitrates on the microbiological safety of meat products. EFSA J 14:1–85

European Food Safety Authority (2007) Opinion of the scientific panel on contaminants in the food chain on a request from the European Commission related to hormone residues in bovine meat and meat products. EFSA J 510:1–62

Elias RJ, Kellerby SS, Decker EA (2008) Antioxidant activity of proteins and peptides. Crit Rev Food Sci Nutr 48:430–441

Elmore JS, Mottram DS, Enser M, Wood JD (2000) The effects of diet and breed on the volatile compounds of cooked meat. Meat Sci 55:149–159

Estévez M, Ollilainen V, Heinonen M (2008) α-Aminoadipic and γ-glutamic semialdehydes as indicators of protein oxidation in myofibrillar proteins. In: Proceedings of the 54th international congress on meat science and technology (ICoMST), Cape Town, South Africa

Estévez M, Morcuende D, Ventanas S (2009) Determination of oxidation. In: Nollet LML, Toldrá F (eds) Handbook of processed meats and poultry analysis. CRC, Boca Raton, FL, pp 221–239

Estévez M (2011) Protein carbonyls in meat systems: a review. Meat Sci 89:259–279

Food and Agriculture Organization/World Health Organization (2008) Benefits and risks of the use of chlorine-containing disinfectants in food production and food processing: report of a joint FAO/WHO expert meeting, Rome, 1–288

Felton JS, Knize MG, Wood C, Wuebbles BJ, Healy SK, Stuermer DH, Bjeldanes LF, Kimble BJ, Hatch FT (1984) Isolation and characterization of new mutagens from fried ground beef. Carcinogenesis 5:95–102

Felton JS, Knize MG, Shen NH, Lewi PR, Anderson BD, Happe J, Hatch FT (1986) The isolation and identification of a new mutagen from fried ground beef: 2-amino-1-methyl-6-phenylimidazol(4,5-b)pyridine. Carcinogenesis 7:1081–1086

Felton JS, Knize MG, Salmon CP, Malfatti MA, Kulp KS (2002) Human exposure to heterocyclic amine food mutagens/carcinogens: relevance to breast cancer. Environ Mol Mutagenes 39:112–118

Fenton SE (2006) Endocrine-disrupting compounds and mammary gland development: early exposure and later life consequences. Endocrinology 147:s18–s24

Ferguson J, Baxter A, Young P, Kennedy G, Elliott C, Weigel S, Gatermann R, Ashwin H, Stead S, Sharman M (2005) Detection of chloramphenicol and chloramphenicol glucuronide residues in poultry muscle, honey, prawn and milk using a surface plasmon resonance biosensor and Qflex® kit chloramphenicol. Anal Chim Acta 529:109–113

Fiddler W, Pensabene JW (1996) Supercritical fluid extraction of volatile N-nitrosamines in fried bacon and its drippings: method comparison. J AOAC Int 79:895–901

Fiems LO, Buts B, Boucque CV, Demeyer DI, Cottyn BG (1990) Effect of a β-agonist on meat quality and myofibrillar protein fragmentation in bulls. Meat Sci 27:29–35

Finley JW, Wheeler EL, Witt SC (1981) Oxidation of glutathione by hydrogen peroxide and other oxidizing agents. J Agric Food Chem 29:404–407

Forte G, Bocca B (2011) Environmental contaminants: heavy metals. In: Nollet LML, Toldrá F (eds) Handbook of analysis of edible animal by-products. CRC, Boca Raton, FL, pp 403–440

Gadgil P, Hachmeister KA, Smith JS, Kropf DH (2002) 2-Alkylcyclobutanones as irradiation dose indicators in irradiated ground beef patties. J Agric Food Chem 50:5746–5750

Gadgil P, Smith JS, Hachmeister KA, Kropf DH (2005) Evaluation of 2-dodecylcyclobutanone as an irradiation dose indicator in fresh irradiated ground beef. J Agric Food Chem 53:1890–1893

García-Regueiro JA, Castellari M (2009) Polychlorinated byphenyls: environmental chemical contaminants. In: Nollet LML, Toldrá F (eds) Handbook of processed meats and poultry analysis. CRC, Boca Raton, FL, pp 635–646

Gatellier P, Kondjoyan A, Portanguen S, Sante-Lhoutelher V (2010) Effect of cooking on protein oxidation in n-3 polyunsaturated fatty acids enriched beef. Implication on nutritional quality. Meat Sci 85:645–650

Gaudin V, Cadieu N, Maris P (2003) Inter-laboratory studies for the evaluation of ELISA kits for the detection of chloramphenicol residues in milk and muscle. Food Agric Immunol 15:143–157

Gentili A, Perret D, Marchese S (2005) Liquid chromatography-tandem drugs in animal-food products. Trends Anal Chem 24:704–733

Gentili A (2007) MS methods for analyzing anti-inflammatory drugs in animal-food products. Trends Anal Chem 26:595–608

Gil L, Barat JM, Garcia-Breijo E, Ibañez J, Martínez-Máñez R, Soto J, Llobet E, Brezmes J, Aristoy MC, Toldrá F (2008) Fish freshness analysis using metallic potentiometric electrodes. Sens Actuators B Chem 131:362–370

Giulivi C, Traaseth NJ, Davies KJA (2003) Tyrosine oxidation products: analysis and biological relevance. Amino Acids 25:227–232

Godfrey MAJ (1998) Immunoafinity extraction in veterinary residue analysis: a regulatory viewpoint. Analyst 123:2501–2506

Gonzalo-Lumbreras R, Izquierdo-Hornillos R (2000) High-performance liquid chromatography optimization study for the separation of natural and synthetic anabolic steroids: application to urine and pharmaceutical samples. J Chromatogr B 742:1–11

Guardiola F, Codony R, Addis PB, Rafecas M, Boatella P (1996) Biological effects of oxysterols: current status. Food Chem Toxicol 34:193–198

Guo JJ, Chou HN, Liao IC (2003) Disposition of 3-(4-cyano-2-oxobutylidene amino)-2-oxazolidone, a cyano-metabolite of furazolidone, in furazolidone-treated grouper. Food Addit Contam 20:229–236

Haasnoot W, Gerçek H, Cazemier G, Nielen MWF (2007) Biosensor immunoassay for flumequine in broiler serum and muscle. Anal Chim Acta 586:312–318

Hagren V, Connolly L, Elliott CT, Lovgren T, Tuomola M (2005) Rapid screening method for halofuginone residues in poultry eggs and liver using time-resolved fluorometry combined with the all-in-one dry chemistry assay concept. Anal Chim Acta 529:21–25

Haughey SA, Baxter GA, Elliot CT, Persson B, Jonson C, Bjurling P (2001) Determination of clenbuterol residues in bovine urine by optical immunobiosensor assay. J AOAC Int 84:1025–1030

Haughey SA, Baxter CA (2006) Biosensor screening for veterinary drug residues in foodstuffs. J AOAC Int 89:862–867

He JH, Hou XL, Jiang HY, Shen JZ (2005) Multiresidue analysis of avermectins in bovine liver by immunoaffinity column cleanup procedure and liquid chromatography with fluorescence detector. J AOAC Int 88:1099–1103

He L, Liu K, Su Y, Zhang J, Liu Y, Zeng Z, Fang B, Zhang G (2011) Simultaneous determination of cyadox and its metabolites in plasma by high-performance liquid chromatography tandem mass spectrometry. J Sep Sci 34:1755–1762

Heggum C (2004) Risk analysis and quantitative risk management. In: Jensen W, Devine C, Dikemann M (eds) Encyclopedia of meat sciences. Elsevier, London, pp 1192–1201

Hernández-Cázares A, Aristoy MC, Toldrá F (2010) Hypoxanthine-based enzymatic sensor for determination of pork meat freshness. Food Chem 123:949–954

Hernández-Cázares AS, Aristoy MC, Toldrá F (2011) An enzyme sensor for the determination of total amines in dry-fermented sausages. J Food Eng 106:166–169

Hernández-Jover T, Izquierdo-Pulido M, Veciana-Nogués MT, Vidal-Carou MC (1996) Biogenic amine sources in cooked cured shoulder pork. J Agric Food Chem 44:3097–3101

Hewitt SA, Kearney M, Currie JW, Young PB, Kennedy DG (2002) Screening and confirmatory strategies for the surveillance of anabolic steroid abuse within Northern Ireland. Anal Chim Acta 473:99–109

Hill LH, Webb NB, Mongol LD, Adams AT (1973) Changes in residual nitrite in sausages and luncheon meat products during storage. J Milk Food Technol 36:515–519

Honikel KO (2009) Oxidative changes and their control in meat and meat products. In: Toldrá F (ed) Safety of meat and processed meat. Springer, Berlin Heidelberg New York, pp 313–340

Honikel KO (2010) Curing. In: Toldrá F (ed) Handbook of meat processing. Blackwell, Ames, IA, pp 125–141

Horvatovich P, Miesch M, Hasselmann C, Delincee H, Marchioni E (2005) Determination of monounsaturated alkyl side chain 2-alkylcyclobutanones in irradiated foods. J Agric Food Chem 53:5836–5841

Hotchkiss JH, Vecchio AL (1985) Nitrosamines in fired-out bacon fat and its use as a cooking oil. Food Technol 39:67–73

Hotchkiss JH, Parker RS (1990) Toxic compounds produced during cooking and meat processing. In: Pearson AM, Dutson TR (eds) Meat and health. Elsevier, London, pp 105–134

Hu Y, Li Y, Liu R, Tan W, Li G. 2011. Magnetic molecularly imprinted polymer beads prepared by microwave heating for selective enrichment of β-agonists in pork and pig liver samples. Talanta 84:462–470

Huang L, Tao YWY, Chen D, Yuan Z (2008) Development of high performance liquid chromatographic methods for the determination of cyadox and its metabolites in plasma and tissues of chicken. J Chromatogr B 874:7–14

Huet AC, Mortier L, Daeseleire E, Fodey T, Elliott CT, Delahaut P (2005) Development of an ELISA screening test for nitroimidazoles in egg and chicken muscle Anal Chim Acta 534: 157–162

Iamiceli AL, Fochi I, Brambilla G, Di Domenico A (2009) Determination of persistent organic pollutants in meat. In: Nollet LML, Toldrá F (eds) Handbook of processed meats and poultry analysis. CRC, Boca Raton, FL, pp 789–836

Jaksyn P, Agudo A, Ibañez R, García-Closas R, Pera G, Amiano P, González CA (2004) Development of a food database of nitrosamines, heterocyclic amines, and polyccyclic aromatic hydrocarbons. J Nutr 134:2011–2014

Jennings WG (1990) Analysis of liquid smoke and smoked meat volatiles by headspace gas chromatography. Food Chem 37:135–144

Kanner J (1994) Oxidative processes in meat and meat products: quality implications. Meat Sci 36:169–189

Kaufmann A (2009) Validation of multiresidue methods for veterinary drugs residues; related problems and possible solutions. Anal Chim Acta 637:144–155

Kaufmann A, Butcher P, Maden K, Walker S, Widmer M (2011) Development of an improved high resolution mass spectrometry based multi-residue method for veterinary drugs in various food matrices. Anal Chim Acta 700:86–94

Kinsella B, O'Mahony J, Malone E, Moloney M, Cantwell H, Furey A, Danaher M (2009) Current trends in sample preparation for growth promoter and veterinary drug residue analysis. J Chromatogr A 1216:7977–8015

Kirbis A, Marinsek J, Flajs VC (2005) Introduction of the HPLC method for the determination of quinolone residues in various muscle tissues. Biomed Chromatogr 19:259–265

Koole A, Franke J-P, De Zeeuw RA (1999) Multi-residue analysis of anabolics in calf urine using high-performance liquid chromatography with diode-array detection. J Chromatogr B 724: 41–51

Kootstra PR, Kuijpers CJPF, Wubs KL, van Doorn D, Sterk SS, van Ginkel LA, Stephany RW (2005) The analysis of beta-agonists in bovine muscle using molecular imprinted polymers with ion trap LCMS screening. Anal Chim Acta 529:75–81

Kumar K, Thompson A, Singh AK, Chander Y, Gupta SC (2004) Enzyme-linked immunosorbent assay for ultratrace determination of antibiotics in aqueous samples. J Environ Qual 33:250–256

Le Bizec B, Pinel G, Antignac J (2009) Options for veterinary drug analysis using mass spectrometry. J Chromatogr A 1216:8016–8034

Lee HJ, Lee MH, Ryu PD, Lee H, Cho MH (2001) Enzyme-linked immunosorbent assay for screening the plasma residues of tetracycline antibiotics in pigs. J Vet Med 63:553–556
Leffers H, Naesby M, Vendelbo B, Skakkebaek NE, Jorgensen M. 2001. Oestrogenic potencies of zeranol, oestradiol, diethylstilboestrol, bisphenol A and genistein: implications for exposure assessment of potential endocrine disrupters. Human Reproductivity 16, 1037–1045.
Levieux D (2007) Immunodiagnosctic technology and its applications. In: Nolle LML, Toldrá F (eds) Advances in food diagnostics. Blackwell, Ames, IA, pp 211–227
Li Y, Slavik MF, Walker JT, Xiong H (1997) Pre-chill spray of chicken carcasses to reduce Salmonella typhimurium. J Food Sci 62:605–607
Link N, Weber W, Fussenegger M (2007) A novel generic dipstick-based technology for rapid and precise detection of tetracycline, streptogramin and macrolide antibiotics in food samples. J Biotechnol 128:668–680
Lipton SH, Bodwell CE, Coleman AH Jr (1977) Amino acid analyzer studies of the products of peroxide oxidation of cystine, lanthionine and homocystine. J Agric Food Chem 25:624–628
Lone KP (1997) Natural sex steroids and their xenobiotic analogs in animal production: growth, carcass quality, pharmacokinetics, metabolism, mode of action, residues, methods, and epidemiology. Crit Rev Food Sci Nutr 37:93–209
Maga JA (1987) The flavour chemistry of wood smoke. Food Rev Int 3:139–183
Marchioni E, Horvatovich P, Ndiaye B, Miesch M, Hasselmann C (2002) Detection of low amount of irradiated ingredients in non-irradiated precooked meals. Radiat Phys Chem 63:447–450
McGlinchey TA, Rafter PA, Regan F, McMahon GP (2008) A review of analytical methods for the determination of aminoglycoside and macrolide residues in food matrices. Anal Chim Acta 624:1–15
McGrath T, Baxter A, Ferguson J, Haughey S, Bjurling P (2005) Multi-sulfonamide screening in porcine muscle using a surface plasmon resonance biosensor. Anal Chim Acta 529:123–127
Maurer HH, Tenberken O, Kratzsch C, Weber AA, Peters FT (2004) Screening for library-assisted identification and fully validated quantification of 22 beta-blockers in blood plasma by liquid chromatography-mass spectrometry with atmospheric pressure chemical ionization. J Chromatogr, 1058, 169–181
Micha R, Wallace SK, Mozaffarian D (2010) Red and processed meat consumption and risk of incident coronary heart disease, stroke, and diabetes mellitus: a systematic review and meta-analysis. Circulation 121:2271–2283
Miller LF, Judge MD, Dikeman MA, Hudgens RE, Aberle ED (1989) Relationships among intramuscular collagen, serum hydroxyproline and serum testosterone in growing rams and wethers. J Anim Sci 67:698–703
Miller LF, Judge MD, Schanbacher BD (1990) Intramuscular collagen and serum hydroxyproline as related to implanted testosterone and estradiol 17β in growing wethers. J Anim Sci 68:1044–1048
Moats WA (1994) Chemical residues in muscle foods. In: Muscle foods: meat, poultry and seafood technology. Kinsman DM, Kotula AW, Breidenstein BC (eds) pp 288–295, New York: Chapman and Hall.
Monarca S, Rizzoni M, Gustavino B, Zani C, Alberti A, Feretti D, Zerbini I (2003) Genotoxicity of surface water treated with different disinfectants using in situ plant tests. Environ Mol Mutagenes 41:353–359
Monarca S, Zani C, Richardson SD, Thruston AD Jr, Moretti M, Feretti D, Villarini M (2004) A new approach to evaluating the toxicity and genotoxicity of disinfected drinking water. Water Res 38:3809–3819
Monsón F, Sañudo C, Bianchi G, Alberti P, Herrera A, Arino A (2007) Carcass and meat quality of yearling bulls as affected by the use of clenbuterol and steroid hormones combined with dexamethasone. J Muscle Foods 18:173–185
Moore WEC, Moore LH (1995) Intestinal floras of populations that have a high risk of colon cancer. Appl Environ Microbiol 61:3202–3207
Mora L, Sentandreu MA, Toldrá F (2008a) Contents of creatine, creatinine and carnosine in pork muscles of different metabolic type. Meat Sci 79:709–715

Mora L, Sentandreu MA, Toldrá F (2008b) Effect of cooking conditions on creatinine formation in cooked ham. J Agric Food Chem 56:11279–11284
Morzel M, Gatellier P, Sayd T, Renerre M, Laville E (2006) Chemical oxidation decreases proteolytic susceptibility of skeletal muscle myofibrillar proteins. Meat Sci 73:536–543
Mottier P, Parisod V, Gremaud E, Guy PA, Stadler RH (2003) Determination of the antibiotic chloramphenicol in meat and seafood products by liquid chromatography–electrospray ionization tandem mass spectrometry. J Chromatogr A 994:75–84
Negri E, Bosetti C, Fattore E, La Vecchia C (2003) Environmental exposure to polychlorinated biphenyls (PCBs) and breast cancer: a systematic review of the epidemiological evidence. Eur J Cancer Prev 12:509–516
Nielsen SS (2010) United States government regulations and international standards related to food analysis 15. In: Nielsen SS (ed) Food analysis, 4th edn. Springer, Berlin Heidelberg New York, pp 15–33
Nout MJR (1994) Fermented foods and food safety. Food Res Int 27:291–296
Oliver CN, Ahn B-W-, Moerman EJ, Goldstein S, Stadtman ER (1987) Age-related changes in oxidized proteins. J Biol Chem 262:5488–5491
Ooizumi T, Xiong YL (2004) Biochemical susceptibility of myosin in chicken myofibrils subjected to hydroxyl radical oxidizing systems. J Agric Food Chem 52:4303–4307
Patsias A, Chouliara I, Paleologos EK, Savvaidis I, Kontominas MG (2006) Relation of biogenic amines to microbial and sensory changes of precooked chicken meat stored aerobically and under modified atmosphere packaging at 4 degrees C. Eur Food Res Technol 223:683–689
Pecorelli I, Bibi R, Fioroni L, Galarini R (2004) Validation of a confirmatory method for the determination of sulphonamides in muscle according to the European Union regulation 2002/657/EC. J Chromatogr A 1032:23–29
Pegg RB, Shahidi F (2000) Nitrite curing of meat. Food & Nutrition, Trumbull, CT, pp 175–208
Peippo P, Lovgren T, Tuomola M. (2005) Rapid screening of narasin residues in poultry plasma by time-resolved fluoroimmunoassay. Anal Chim Acta 529:27–31
Peng Z, Bang-Ce Y (2006) Small molecule microarrays for drug residue detection in foodstuffs. J Agric Food Chem 54:6978–6983
Perry GA, Welshons WV, Bott RC, Smith MF (2005) Basis of melengestrol acetate action as a progestin. Domest Anim Endocrinol 28:147–161
Puente ML (2004) Highly sensitive and rapid normal-phase chiral screen using high-performance liquid chromatography-atmospheric pressure ionization tandem mass spectrometry (HPLC/MS). J Chromatogr 1055:55–62
Ramarathnam N (1998) The flavour of cured meat. In: Shahidi F (ed) Flavor of meat, meat products and seafood. Blackie Academic & Professional, London, pp 290–319
Raoul S, Gremaud E, Biaudet H, Turesky RJ (1997) Rapid solid-phase extraction method for the detection of volatile nitrosamines in food. J Agric Food Chem 45:4706–4713
Rath S, Reyes FG (2009) Nitrosamines. In: Nollet LML, Toldrá F (eds) Handbook of processed meats and poultry analysis. CRC, Boca Raton, FL, pp 687–705
Reeves PT (2007) Residues of veterinary drugs at injection sites. J Vet Pharmacol Ther 30:1–17
Reeves PT (2010) Drug residues. In: Cunningham F, Elliott J, Lees P (eds) Comparative and veterinary pharmacology (Handbook of experimental pharmacology), vol 199. Springer, Berlin Heidelberg New York, pp 265–290
Reig M, Toldrá F (2007) Chemical origin toxic compounds. In: Toldrá F, Hui YH, Astiasarán I, Nip WK, Sebranek JG, Silveira ETF, Stahnke LH, Talon R (eds) Handbook of fermented meat and poultry, Blackwell, Ames, IA, pp 469–475
Reig M, Toldrá F (2008a) Veterinary drug residues in meat: concerns and rapid methods for detection. Meat Sci 78:60–67
Reig M, Toldrá F (2008b) Immunology-based techniques for the detection of veterinary drugs residues in foods. In: Toldrá F (ed) Meat biotechnology. Springer, Berlin Heidelberg New York, pp 361–373

Reig M, Toldrá F (2009a) Veterinary drugs and growth promoters residues in meat and processed meats. In: Toldrá F (ed) Safety of meat and processed meat. Springer, Berlin Heidelberg New York, pp 365–390

Reig M, Toldrá F (2009b) Growth promoters. In: Nollet LML, Toldrá F (eds) Handbook of muscle foods analysis. CRC, Boca Raton, FL, pp 837–854

Reig M, Toldrá F (2009c) Veterinary drug residues. In: Nollet LML, Toldrá F (eds) Handbook of processed meats and poultry analysis. CRC, Boca Raton, FL, pp 647–664

Reig M, Toldrá F (2010) Detection of chemical hazards. In: Toldrá F (ed) Handbook of meat processing. Blackwell, Ames, IA, pp 469–480

Reig M, Toldrá F (2011) Growth promoters. In: Nollet LML, Toldrá F (eds) Safety analysis of foods of animal origin. CRC, Boca Raton, FL, pp 229–247

Reig M, Batlle N, Navarro JL, Toldrá F (2005) A modified HPLC method for the detection of 6-methyl-2-thiouracil in cattle urine. In: Proceedings of the international congress in meat science and technology, Baltimore, MD, 7–12 August 2005

Reig M, Mora L, Navarro JL, Toldrá F (2006) A chromatography method for the screening and confirmatory detection of dexamethasone. Meat Sci 74:676–680

Roda A, Manetta AC, Portanti O, Mirasoli M, Guardigli M, Pasini P, Lelli R (2003) A rapid and sensitive 384-well microtitre format chemiluminiscent enzyme immunoassay for 19-nortestosterone. Luminescence 18(2):72–78

Ross CF, Smith DM (2006) Use of volatiles as indicators of lipid oxidation in muscle foods. Compr Rev Food Sci Saf 5:18–25

Rowe LJ, Maddock KR, Lonergan SM, Huff-Lonergan E (2004a) Influence of early postmortem protein oxidation on beef quality. J Anim Sci 82:785–793

Rowe LJ, Maddock KR, Lonergan SM, Huff-Lonergan E (2004b) Oxidative environments decrease tenderization of beef steaks through inactivation of l-calpain. J Anim Sci 82:3254–3266

Ruiz-Capillas C, Jiménez-Colmenero F (2010) Biogenic amines in seafood products. In: Nollet LML, Toldrá F (eds) Handbook of seafood and seafood products analysis. CRC, Boca Raton, FL, pp 833–850

Santarelli RL, Vendeuvre J-L, Naud N, Taché S, Guéraud F, Viau M, Genot C, Corpet DE, Pierre FFH (2010) Meat processing and colon carcinogenesis: cooked nitrite-treated and oxidized high-heme cured meat promotes mucin-depleted foci in rats. Cancer Prev Res 3:852–864

Scientific Committee for Food (1995) Smoke flavorings. Report of the Scientific Committee for Food of the European Commission. Opinion adopted on 23 June 1993. Series 34: Food Science Techniques

Sen NP, Baddoo PA, Seaman SW (1987) Volatile nitrosamines in cured meats packaged in elastic rubber nettings. J Agric Food Chem 35:346–350

Shalaby AR (1996) Significance of biogenic amines to food safety and human health. Food Res Int 29:675–690

Shao B, Jia X, Zhang J, Meng J, Wu Y, Duan H, Tu X (2009) Multi-residual analysis of 16 β-agonists in pig liver, kidney and muscle by ultra performance liquid chromatography tandem mass spectrometry. Food Chem 114:1115–1121

Shi WM, He JH, Jiang HY, Hou XL, Yang JH, Shen JZ (2006) Determination of multiresidue of avermectins in bovine liver by an indirect competitive ELISA. J Agric Food Chem 54: 6143–6146

Sikorski ZE, Kolakowski E (2010) Smoking. In: Toldrá F (ed) Handbook of meat processing. Blackwell, Ames, IA, pp 231–245

Simko P (2009a) Polycyclic aromatic hydrocarbons in smoked meats. In: Toldrá F (ed) Safety of meat and processed meat. Springer, Berlin Heidelberg New York, pp 343–363

Simko P. 2009b. Polycyclic aromatic hydrocarbons. In: Nollet LML, Toldrá F (eds) Handbook of processed meats and poultry analysis. CRC, Boca Raton, FL, pp 707–724

Sinha R, Rothman N, Salmon CP, Knize MG, Brown ED, Swanson CA, Rhodes D, Rossi S, Felton JS and Levander OA (1998) Heterocyclic amine content in beef cooked by different methods to varying degrees of doneness and gravy made from beef drippings. Food Chem Toxicol 36:279–287

Situ C, Elliott CT (2005) Simultaneous and rapid detection of five banned antibiotic growth promoters by immunoassay. Anal Chim Acta 529:89–96

Situ C, Grutters E, van Wichen P, Elliott CT (2006) A collaborative trial to evaluate the performance of a multi-antibiotic enzyme-linked immunosorbent assay for screening five banned antimicrobial growth promoters in animal feedingstuffs. Anal Chim Acta 561:62–68

Slump P, Schreuder HAW (1973) Oxidation of methionine and cystine in foods treated with hydrogen peroxide. J Sci Food Agric 24:657–661

Smith JS, Pillai S (2004) Irradiation and food safety. Food Technol 58:48–55

Sommers CH, Delinceé H, Smith JS, Marchioni E (2006) Toxicological safety of irradiated foods. In: Sommers CH, Fan X (eds) Food irradiation: research and technology, Blackwell, Ames, IA, pp 1–55

Srinivasan, S, Xiong YL (1996) Gelation of beef heart surimi as affected by antioxidants. J Food Sci 61:707–711

Srinivasan S, Hultin HO (1997) Chemical, physical, and functional properties of cod proteins modified by a nonenzymic free-radical-generating system. J Agric Food Chem 45:310–320

Stefanova R, Vasilev NV, Spassov SL (2010) Irradiation of food, current legislation framework, and detection of irradiated foods. Food Anal Methods 3:225–252

Stewart EM, McRoberts WC, Hamilton JTG, Graham WD (2001) Isolation of lipid and 2-alkylcyclobutanones from irradiated foods by supercritical fluid extraction. J AOAC Int 84:976–986

Stewart EM (2009) Detection of irradiated ingredients. In: Nollet LML, Toldrá F (eds) Handbook of processed meats and poultry analysis. CRC, Boca Raton, FL, pp 725–745

Stolker AAM, Schwillens P-L-WJ, van Ginkel LA, Brinkman UATh (2000) Comparison of different liquid chromatography methods for the determination of corticosteroids in biological matrices. J Chromatogr A 893:55–67

Straub BW, Kicherer M, Schilcher SM, Hammes WP (1995) The formation of biogenic amines by fermentation organisms. Z Lebensm Unters Forsch 201:79–82

Stubbings G, Tarbin J, Cooper A, Sharman M, Bigwood T, Robb P (2005) A multi-residue cation-exchange clean up procedure for basic drugs in produce of animal origin. Anal Chim Acta 547: 262–268

Takemura H, Shim JY, Sayama K, Tsubura A, Zhu BT, Shimoi K (2007) Characterization of the estrogenic activities of zearalenone and zeranol in vivo and in vitro. J Steroid BioChem Mol Biol 103:170–177

Talon R, Leroy-Sétrin S, Fadda S (2002) Bacterial starters involved in the quality of fermented meat products. In: Toldrá F (ed) Research advances in the quality of meat and meat products. Research Signpost, Trivandrum, India, pp 175–191

Tewfik I (2008a) Extraction and identification of cyclobutanones from irradiated cheese employing a rapid direct solvent extraction method. Int J Food Sci Nutr 59:590–598

Tewfik I (2008b) Inter-laboratory trial to validate the direct solvent extraction method for the identification of 2-dodecylcyclobutanone in irradiated chicken and whole liquid egg. Food Sci Technol Int 14:277–283

Thevis M, Opfermann G, Schänzer W (2003) Liquid chromatography/electrospray ionization tandem mass spectrometric screening and confirmation methods for beta2-agonists in human or equine urine. J Mass Spectrom 38:1197–1206

Thompson CS, Haughey SA, Traynor IM, Fodey TL, Elliot CT, Antignac J-P, Le Bizec B, Crooks SRH (2008) Effective monitoring of ractopamine residues in samples of animal origin by SPR biosensor and mass spectrometry. Anal Chim Acta 608:217–225

Toldrá F (2004) Fermented meats. In: Hui YH, Smith JS (eds) Food processing: principles and applications. Blackwell, Ames, IA, pp 399–415

Toldrá F (2006a) Biochemistry of fermented meat. In: Hui YH, Nip WK, Nollet ML, Paliyath G, Simpson BK (eds) Food biochemistry and food processing. Blackwell, Ames, IA, pp 641–658

Toldrá F (2006b) Meat fermentation. In: Hui YH, Castell-Perez E, Cunha LM, Guerrero-Legarreta I, Liang HH, Lo YM, Marshall DL, Nip WK, Shahidi F, Sherkat F, Winger RJ, Yam KL (eds) Handbook of food science, technology and engineering. CRC, Boca Raton, FL, vol 4, pp 181–1 to 181–12

Toldrá F, Reig M (2006) Methods for rapid detection of chemical and veterinary drug residues in animal foods. Trends Food Sci Technol 17:482–489

Toldrá F, Reig M (2007) Chemical origin toxic compounds. In: Toldrá F, Hui YH, Astiasarán I, Nip WK, Sebranek JG, Silveira ETF, Stahnke LH, Talon R (eds) Handbook of fermented meat and poultry. Wiley-Blackwell, Ames, IA, pp 469–475

Toldrá F, Aristoy MC, Flores M (2009) Relevance of nitrate and nitrite in dry-cured ham and their effects on aroma development. Grasas y Aceites 60:291–296

Toldrá F, Reig M (2011) Innovations for healthier processed meats. Trends Food Sci Technol 22:517–522

Toldrá F, Reig M (2012) Residue analysis. In: Jensen W, Devine C, Dikemann M (eds) Encyclopedia of meat sciences, 2nd edn. Elsevier, London (in press)

Tsai LS, Wilson R, Randall V (1995) Disinfection of poultry chilled water with chlorine dioxide: consumption and by-product formation. J Agric Food Chem 43:2768–2773

Twaroski TP, O'Brien ML, Robertson LW (2001) Effects of selected polychlorinated biphenyl (PCB) congeners on hepatic glutathione, glutathione-related enzymes and selenium status: implications for oxidative stress. Biochem Pharmacol 62:273–281

United States Department of Agriculture (2002a) The use of chlorine dioxide as an antimicrobial agent in poultry processing in the United States. USDA-FSIS, Office of International Affairs, Washington, DC, November 2002

United States Department of Agriculture (2002b) The use of acidified sodium chlorite as an antimicrobial agent in poultry processing in the United States. USDA-FSIS, Office of International Affairs, Washington, DC, December 2002

United States Department of Agriculture (2002c) The use of trisodium phosphate as an antimicrobial agent in poultry processing in the United States. USDA-FSIS, Office of International Affairs, Washington, DC, November 2002

United States Department of Agriculture (2002d) The use of peroxyacids as an antimicrobial agent in poultry processing in the United States. USDA-FSIS, Office of International Affairs, Washington, DC, December 2002

United States Food and Drug Administration (1994) Pesticide analytical manual, 3rd edn. Volume I updated in October 1999 and Volume II updated in January 2002. National Technical Information Service, Springfield, VA. http://www.fda.gov/Food/ScienceResearch/LaboratoryMethods/PesticideAnalysisManualPAM/ucm111455.htm. (Accessed 30 Mar 2012)

Van Bocxlaer JF, Casteele SRV, Van Poucke CJ, Van Peteghem CH (2005) Confirmation of the identity of residues using quadrupole time-of-flight mass spectrometry. Anal Chim Acta 529: 65–73

Van den Bogaard, AEJM, London N, Stobberingh EE (2000) Antimicrobial resistance in pig faecal samples from The Netherlands (five abattoirs) and Sweden. J Antimicrob Chem 45:663–671

Van der Heeft E, Bolck YJC, Beumer B, Nijrolder AWJM, Stolker AAM, Nielen MWF (2009) Full-scan accurate mass selectivity of ultra-performance liquid chromatography combined with time-of-flight and orbitrap mass spectrometry in hormone and veterinary drug residue analysis. J Am Soc Mass Spectrom 20:451–463

Van Peteguem C, Daeselaire E (2004) Residues of growth promoters. In: Nollet LML (ed) Handbook of food analysis, 2nd edn. Dekker, New York, pp 1037–1063

Variyar PS, Chatterjee S, Sajilata MG, Singhal RS, Sharma A (2008) Natural existence of 2-alkylcyclobutanones. J Agric Food Chem 56:11817–11823

Vázquez-Roig, Picó Y (2011) Environmental contaminants: pesticides. In: Nollet LML, Toldrá F (eds) Handbook of analysis of edible animal by-products. CRC, Boca Raton, FL, pp 377–402

Verdon E, Couedor P, Roudaut B, Sanders P (2005) Multiresidue method for simultaneous determination of ten quinolone antibacterial residues in multimatrix/multispecies animal tissues by liquid chromatography with fluorescence detection: single laboratory validation study. J AOAC Int 88:1179–1192

Verdon E (2008) Antibiotic residues in muscle tissues of edible animal products. In: Nollet LML, Toldrá F (eds) Handbook of meat products analysis. CRC, Boca Raton, FL, pp 856–947

Vidal-Carou MC, Veciana-Nogués M, Latorre-Moratalla ML, Bover-Cid S (2007) Biogenic amines: risk and control. In: Toldrá F, Hui YH, Astiasarán I, Nip WK, Sebranek JG, Silveira ETF, Stahnke LH, Talon R (eds) Handbook of fermented meat and poultry. Wiley-Blackwell, Ames, IA, pp 455–468

Vidal-Carou MC, Latorre-Moratalla ML, Bover-Cid S (2009) Biogenic amines. In: Nollet LML, Toldrá F (eds) Handbook of processed meats and poultry analysis. CRC, Boca Raton, FL, pp 665–686

Viljanen K, Kylli P, Kivikari R, Heinonen M (2004) Inhibition of protein and lipid oxidation in liposomes by berry phenolics. J Agric Food Chem 52:7419–7424

Vinci G, Antonelli ML (2002) Biogenic amines: quality index of freshness in red and white meat. Food Control 13:519–524

Vollard EJ, Clasener HAL (1994) Colonization resistance. Antimicrob Agents Chemother 38: 409–414

Walker R (1990) Nitrates, nitrites and nitrosocompounds: a review of the occurrence in food and diet and the toxicological implications. Food Addit Contam 7:717–768

Wang S, Wang ZL, Duan ZJ, Kennedy I (2006) Analysis of sulphonamide residues in edible animal products: a review. Food Addit Contam 23:362–384

Wang S, Wang XH (2007) Analytical methods for the determination of zeranol residues in animal products: a review. Food Addit Contam 24:573–582

Wenzl T, Simon R, Kleiner J, Anklam E (2006) Analytical methods for polycyclic aromatic hydrocarbons (PAHs) in food and the environment needed for new food legislation in the European Union. Trends Anal Chem 25:716–725

Widstrand C, Larsson F, Fiori M, Civitareale C, Mirante S, Brambilla G (2004) Evaluation of MISPE for the multi-residue extraction of beta-agonists from calves urine. J Chromatogr B 804:85–91

Wilson VS, Lambright C, Ostby J, Gray LE Jr (2002) In vitro and in vivo effects of 17 beta-trenbolone: a feedlot effluent contaminant. Toxicol Sci 70:202–211

Xiong YL (2000) Protein oxidation and implications for muscle food quality. In: Decker EA, Faustman C, López-Bote CJ (eds) Antioxidants in muscle foods. Wiley, New York, pp 85–111

Xu CL, Chu XG, Peng CF, Liu LQ, Wang LY, Jin Z (2006a) Comparison of enzyme-linked immunosorbent assay with liquid chromatography-tandem mass spectrometry for the determination of diethylstilbesterol residues in chicken and liver tissues. Biomed Chromatogr 20:1956–1064

Xu CL, Peng CF, Liu LQ, Wang LY, Jin ZY, Chu XG (2006b) Determination of hexoestrol residues in animal tissues based on enzyme-linked immunosorbent assay and comparison with liquid chromatography-tandem mass spectrometry. J Pharm Biomed Anal 41:1029–1036

Zanardi E, Battaglia A, Ghidini S, Conter M, Badiani A, Ianieri A (2007) Evaluation of 2-alkylcyclobutanones in irradiated cured pork products during vacuum-packed storage. J Agric Food Chem 55:4264–4270

Zhang YL, Huang LL, Chen DM, Fan SX, Wang YL, Tao YF, Yuan ZH (2005) Development of HPLC methods for the determination of cyadox and its main metabolites in goat tissues. Anal Sci 21:1495–1499

Zhang W, Wang HH, Wang JP, Li XW, Jiang HY, Shen JZ (2006a) Multiresidue determination of zeranol and related compounds in bovine muscle by gas chromatography/mass spectrometry with immunoaffinity cleanup. J AOAC Int 89:1677–1681

Zhang SX, Zhang Z, Shi WM, Eremin SA, Shen JZ (2006b) Development of a chemiluminescent ELISA for determining chloramphenicol in chicken muscle. J Agric Food Chem 54: 5718–5722

Index

F. Toldrá and M. Reig, *Analytical Tools for Assessing the Chemical Safety of Meat and Poultry*, SpringerBriefs in Food, Health, and Nutrition 9,
DOI 10.1007/978-1-4614-4277-6,